Generis

PUBLISHING

A finite volume numerical method for solving 2D conduction in a rectangular plate using MATLAB

BENOUAZ Khawla

Author: BENOUAZ Khawla

ISBN: 978-1-63902-171-0

Title: A finite volume numerical method for solving 2D conduction in a rectangular plate using MATLAB

Cover image: www.unsplash.com

Generis Publishing

Online orders: www.generis-publishing.com

Orders by email: info@generis-publishing.com

"....."بِسْمِ اللَّـهِ الرَّحْمَـٰنِ الرَّحِيمِ..."

"... In the Name of Allah, the Beneficent, the Merciful..."

"Of course, there is hard work,

But, has the joy of success

Not to compensate for our pain? "

Jean de la Bruyere

I dedicate this work to my parents

KHAWLA

Table of Contents

List of Figures

List of paintings

NOMENCLATURE

A- List of Latin symbols:

Symbols	Designations	Units
T	Dimensional Temperature	$[K]$
A	Object surface	$[m^2]$
Cp	Specific heat at constant pressure	$[J.kg^{-1}.K^{-1}]$
Q	The amount of heat	$[W]$
t	Time	$[s]$
k	Thermal conductivity	$[W.m^{-1}.K^{-1}]$
h	Convection coefficient	$[W.m^{-2}.K^{-1}]$
g	Acceleration of gravity	$[m/s^2]$
U	Electric tension	$[V]$
I	Electric intensity	$[A]$
R	Electrical resistance	$[\Omega]$
R_t	Thermal resistance	$[KW^{-1}]$
D	Characteristic dimension of the exchange surface	$[m]$
U_m	Average fluid velocity	$[m.s^{-1}]$
x, y	Cartesian coordinates of the system	$[m]$
H	plate height	$[m]$
L	The length of the plate	$[m]$
e	Thickness of the object	$[m]$
C	speed of light in a vacuum	$[m.s^{-1}]$

$\dfrac{\partial}{\partial x}$	Partial derivative	$[-]$
$\dfrac{d}{dx}$	Total derivative	$[-]$
$\vec{n}$	Unit vector	$[-]$
dS	elementary surface	$[m]$
L	Global energy luminance	$\left[w.m^{-2}sr^{-1} \right]$
L_λ	Monochromatic luminance	$\left[w.m^{-2}sr^{-1}.m^{-1} \right]$
M	Monochromatic energy emittance	$\left[w.m^{-2}.m^{-1} \right]$
M_λ	Global energy emittance	$\left[w.m^{-2} \right]$
gradT	temperature gradient	$\left[K.m^{-1} \right]$
diV	Divergence	$[-]$
[A]	Matrix	$[-]$
$\{U\}$	Unknown vectors	$[-]$
$\{L\}$	Vector	$[-]$

B- List of Greek symbols :

Symbols	designations	Units
ΔT	Temperature difference	$[K]$
Δx	Dimension of a control volume in the horizontal direction	$[m]$
Δy	Dimension of a control volume in the vertical direction	$[m]$
θ	Dimensionless temperature	$[K]$

λ	Wave length	$[m]$
θ_i	Indoor temperature	$[K]$
θ_e	Outdoor temperature	$[K]$
Φ	Heat flux	$[W]$
φ	Heat flux density	$[W/m^2]$
ρ	Volumic mass	$[kg.m^{-3}]$
Σ	Contour	$[-]$
μ	Dynamic viscosity	$[kg.m^{-1}s^{-1}]$
ε	Form factor between surfaces i and j	$[-]$
σ	STEFAN-BOLTZMAN constant	$[W/m^2K^4]$
Ω	Solid angle	$[sr]$

C- Dimensionless Numbers

Symbols	Designations	Units
Gr	Number of Grashof	$[-]$
Ra	Rayleigh number	$[-]$
Nu	Nombre de Nusselt local	$[-]$
Pr	Number of Prandtl	$[-]$
Pe	Number of Péclet	$[-]$

D- Lower Indices

Symbols	Designations	Units
i, j	i, j i th and j th components	$[-]$

P	Point at the center of the finished volume	$[-]$
E	Point east of point P	$[-]$
W	Point west of point P	$[-]$
N	Point north of point P	$[-]$
S	Point south of point P	$[-]$
e	East face of the considered control volume	$[-]$
w	West face of the considered control volume	$[-]$
n	North face of the considered control volume	$[-]$
s	South face of the considered control volume	$[-]$

E- **Abbreviations**

Abbreviations	**designations**
FDM	Finite Difference Method
FVM	Finite volume method
FEM	Finite element method
EDP	Partial derivative equations
TDMA	Tridiagonal Matrix Algorithm

General Introduction

General introduction

The three major numerical methods used in three-dimensional codes are the finite volumes, finite differences and finite elements.

The finite volume method, used by several commercial codes, and open source such as

ANSYS CFX, ANSYS Fluent, OpenFoam and others software is widely described by **PATANKAR**, it consists in discretizing the field of flow in a multitude of control volumes (cells) and then perform balances (mass, quantity of movement ...) on these small volumes. For this reason, the wording shows triple volume integrals.

This method is used to solve derivative equations numerically partial , like other numerical methods.

In the present work, we study the finite volume method to solve the conduction 2D in a rectangular plate using the **MATLAB** calculation code.

To achieve these objectives, the work presented in this thesis is organized into five chapters:

After this brief introduction, we discuss, in:

The first chapter: bibliographic research.

A bibliographical summary was presented on the resolution of the governing equation by numerical methods.

The second chapter: general information on heat transfer modes a brief idea on the basics of heat transfer and generalities on the three modes heat transfer.

The third chapter: generalities on the methods of resolution a reminder on the three main families of resolution methods.

The fourth chapter: model of the governing equation.

Is interested in the numerical discretization of the governing equation by the method of finished volumes.

The fifth chapter: result and interpretation

This is the most important part of this thesis, it includes the presentation of our study on the MATLAB calculation code, as well as the interpretation and validation of the results obtained.

Finally, our brief ends with a general conclusion.

Chapter I:

Bibliographic search

I. Chapter I. Bibliographic research

I.1 Introduction:

There are many ways to represent continuous problems in a discrete fashion. as for example the approximations by finite differences, by finite elements, or by finished volumes.

After this preface, we present some available works which deal with the resolution of governing equations by the three methods.

I.2 Review of previous studies:

M.T. Manzari et M.T.Manzari 1999 [1] , carried out a study on the digital resolution hyperbolic equation of heat conduction by the finite element method, the system of equations is solved for temperature and heat fluxes as variables independent.

The standard Galerkin method is used for spatial discretization and the Crank-method Nicolson is adopted for walking in the time domain. They showed that the method proposed can easily assess the entropy production in the domain and estimate the thermodynamic equilibrium of the system. The performance of the proposed algorithm is verified by solving a 1D and 2D test case, and some interesting characteristics of the hyperbolic heat conduction have been demonstrated.

A. Grine, J.Y.Desmons Et S.Harmand 2006 [2], conducted a study on models for transient conduction in a protective plate subjected to a variable heat flow, they determined the development of analytical models for the identification of evolution of the temperature distribution in a plate with respect to time.

Analytical models describing the temperature distribution in a plate exposed to heat flow, Green's method is used for the development of these models. The results showed the evolution of the temperature compared to the number of Biot and Fourier.

A. G.Hansen, M. P. Bendsoe Et O. Sigmund 2006[3], Carried out a study on the use of the finite volume method to solve an optimization problem topological prototype, they considered a heat diffusion problem for the resolution.

They used the application of FVM to problems with distributions of materials not homogeneous, to provide a single value for the boundary conductivity of the elements are arithmetic and harmonic means were used.

They observed that when using the harmonic mean, the checkerboards do not form during the topology optimization process.

They have shown that topology optimization is possible in the volume method finished.

A.V. Itagi 2007 [4], conducted a study on the finite volume method to analyze the thermal transport in anisotropic layered media in the presence of a volumetric thermal source.

The method used to obtain the steady state as well as the transient temperature profiles is the implicit Douglas-Gunn method in alternating direction, the method used solved the problem in the reference frame of the source and uses a non-uniform polar grid.

The method shows excellent agreement with the analysis results.

A.Diószegi, Et All 2015[5], presented a Modeling and simulation of conduction thermal in 1D polar spherical coordinates using the finite difference method based on the control volume, they used the CVFDM method.

A mixture of sand and different chemicals (binders) is used as the material for molding in casting processes.

The simulation results were validated by comparison with temperature measurements under laboratory conditions when the sand mold mixture has been heated in interacting with a liquid alloy, the numerical method indicated is exact and presents a significant potential in the simulation of casting processes.

G. Sachdeva, K.S. Kasana, Et R. Vasudevan 2010[6], conducted a numerical study of a laminar flow, incompressible and viscous, in triangular shaped fins with a vortex generator, they used three different angles of attack of the wing to achieve this study, that is, 15 °, 20 ° and 26 °.

The Marker and Cell (MAC) method is used to obtain the components of pressure and speed as well as the resolution of the Navier-Stokes equation.

The results showed that using a rectangular wing vortex generator at a approach angle of 26 ° results in an increase of approximately 35% in the average number of Nusselt medium compared to triangular fin heat exchanger without generator whirlpool.

S. Mazumder 2017[7], carried out a comparative study between the solutions produced by three methods of resolution on a one-dimensional coordinate system for a problem of conduction, they determined the accuracy of the temperature, the accuracy of the heat flow and the satisfaction of global energy conservation.

He concluded the same discretization schemes for the three methods, the precision of the temperature produced is similar for finite differance and finite element but the method of finished volumes is a little different.

X. Xiaofeng Et X. Qiong 2012[8], conducted a study on solving the equation of the two-dimensional conduction by the finite volume method whch is performed by programming MATLAB on an infinite plate of uniform thickness and a two-dimensional rectangle.

The algebraic equation discretized by the finite volume method, different coefficients and source terms have been discussed under different boundary conditions, including heat flux prescribed, prescribed temperature, convection and insulation.

The feasibility and the stability of the numerical method have been demonstrated by the current result.

S. Murakami Et Y. Asako 2011 [9], Conducted a study on quadrilateral meshes deformed, solved by the finite volume method, the method discretizes the conduction term.

In this method, it is possible to compose the calculation mesh of the elements general quadrilateral, The test calculations show that the trend of convergence of the numerical error using this method with the deformed mesh is the same as the use of a diagram of central difference with two nodes on a rectangular mesh with constant interval, the error of numerical results by this method is smaller than the use of the traditional multilateral elements method.

The results show that the present method with the deformed mesh agrees well with the analytical solution and the result of REM with a rectangular mesh.

B. Mondal Et S. C. Mishra 2008[10], carried out a study on the Boltzmann method in network (LBM) in conjunction with the finite volume method for problem solving of conduction radiation combined with temperature as well as conditions at flux limits, on 1D and 2D geometries.

The LBM was used to solve the energy equation, and using the FVM to calculate the radiative information required in the energy equation.

In 1D geometry, the southern boundary is subjected to a constant heat flow, and the geometry

2D the southern and northern limit is at a state of constant heat, the remaining limits are at prescribed temperatures.

They have shown a successful implementation of LBM in conjunction with FVM to a class more general of problems having temperature and flow conditions.

W.Li, B.Yu, Et All 2012[11], Have carried out a study on a new method of volumes finished for cylindrical conduction problems, they determined the difference between the new finite volume method for cylindrical conduction problems and classical second-order methods.

The new method is more precise than the conventional methods, It is noted that this method costs less computation time than that of conventional methods, even if the discretized expression of this proposed method is more complex than the method of the second-order central finite volume.

The numerical result shows that the total time of the new method is less than conventional methods to achieve the same level of precision.

P. Duda 2016 [12], conducted a study to solve heat conduction problems transients by the formulation of the finite element method in the polar coordinates, they showed how to apply the most commonly used boundary conditions.

The Crank-Nicolson method is used for the overall system resolution, he used three different numerical tests to carry out this study, all the examples show the polar coordinate system gives better results than in the polar coordinate system Cartesian coordinates, even the formulation of the finite element method in the polar coordinates is valuable.

This method makes it possible to calculate the temperature distribution in the bodies of different properties in circumferential direction and radial direction.

The tests show good precision and stability of the proposition.

T. Ota 1973[13], carried out a study on the conduction of heat in an infinite plate with a rectangular hole.

The results showed that the two-dimensional distribution of constant heat flow around the rectangular hole in the plate, heat flow to the corner of the rectangle is examined and it is specified that the heat flow possesses.

H. T.Kim, B. W. Rhee Et J. H. Park 2006[14], conducted a study on the applications of radial conduction finite volume method model of the CATHENA code as a example.

They used the digital FEM diagram for the radial conduction model of the code CATHENA replaced by FVM to remove the effect of the mesh size on the fuel temperature prediction.

The finite volume method is applied to the CATHENA wall conduction model for avoid mesh size on fuel temperature prediction, accuracy and validity of the finite volume model in the CATHENA code are tested in two cases, one case stable and a case of transient thermal conduction.

In the stable case they showed one over the boundary surface and a uniform rate of generation of internal heat, in the case of transient thermal conduction, a cylinder is initially at a uniform temperature and suddenly its boundary surface is subjected to convection with a constant heat transfer coefficient in room temperature to temperature constant, the steady-state solutions of the MVF model give almost the same results as analytical solutions.

R. Chaabane, F. Askri Et S. Ben Nasrallah 2011[15], carried out a study on the performance evaluation of the Boltzmann method (LBM) and the method of control volume finite elements (CVFEM).

The effects are studied on temperature distributions, radiative heat flows and conductors, the numerical approach is extended to deal with a practical combination of mixed boundary conditions in a combined radiative heat transfer problem multidimensional transient conductor in an emitting, absorbing, anisotropic diffusion enclosure.

The results of LBM in conjunction with CVFEM were found to compare very well with the results available in the literature.

C. Luo, et all 2009[16], Carried out a study on the modeling of the heat transfer of the wall using modification of the conduction transfer function, finite volume and complex Fourier analysis methods.

The CTF method was used to calculate the heat fluxes of the surface of the walls in depending on the surface temperatures of the walls controlled as inputs, the method of finite volume and matrix method were also used for numeric predictions the conduction transfer function method and the finite volume method were compared to long period measurements for single layer materials or multilayer with and without air gaps.

The CTF coefficients for the modified CTF methods were totaled and analyzed for all the calculation cases in this study.

S. Han 2015[17], carried out a study on the application of the finite volume method for the solution of hyperbolic conduction equations in two steps.

 The method applied for non-Fourier conduction in casting sand, to describe the short-term transient thermal conduction through casting sand when two energy carriers (sand and air) are not in thermal equilibrium. Numerical results show favorable agreement with experimental Existing data.

S.Singh,et all 2007[18], presented a study on transient heat conduction in polar coordinates with several layers in the radial direction.

The variable separation method is used to obtain a distribution of transient temperature, are assumed in each layer of heat sources volumetric spatially non-uniform, but independent of time, the solution proposed is also applicable to several layers with zero internal radius. An example Illustrative problem for the three-layer semicircular annular region is solved. The results with isotherms are presented graphically and discussed.

J. Stefaniak Et J. Jankowski 1998[19], presented a study on the approximate solution heat conduction equation with mixed boundary condition in a rectangular plate.

They used the approximate method of solving linear problems of heat, the method is applicable to direct and reverse problems.

Direct problem, we understand the solution of the heat equation by conduction with given initial limit condition and possibly the source currents of the heat, an inverse problem is to find the intensities of the heat sources, when the temperature of the solid considered is known.

Evgrafov, M.M. Gregersen Et M.P. Sørensen 2011[20], conducted a study on the convergence of a discretization of the cell-based finite volume method for control problems in conduction coefficients.

They applied a control problem in the coefficients of a modeling of the equation de Laplace generalized, for example. They characterized the endpoints of the sequences of discrete global solutions and two types of stationary points, as well as measured the mesh size of the finite volume converges to zero, They noted that this study is a first step towards enabling finite volume discretizations in coefficient problems.

S. Han 2014[21], carried out a study on the solution of the conduction equation 1D hyperbolic by the finite volume method.

The proposed formulation is verified by exact solutions for a homogeneous medium, then applied to composite materials with different conductivity.

The proposed method can be easily extended to heterogeneous two-dimensional media with temperature dependent properties subjected to various boundary conditions.

The results show a very distinct wave penetration and reflection at the interfaces of materials.

I.3   The objective of our study:

The main objective of our thesis is the study and the discretization of the heat equation in 2D in a rectangular plate, using the finite volume method for the resolution with a structured mesh and Cartesian coordinates, with different boundary conditions, using the **MATLAB** language as a programming tool.

Chapter II:

General information on heat transfer modes

II.   Chapter II. General information on heat transfer modes

II.1   Introduction :

Heat transfer is defined by the transmission of energy from one region to another under the influence of a temperature difference.

Heat transfer essentially recognizes three modes of transmission: convection, radiation and conduction. [22]

Each of these modes is itself linked to a well-determined physical process, and are governed by very specific laws.

In this chapter, we explain the mechanism of three modes of heat transfer.

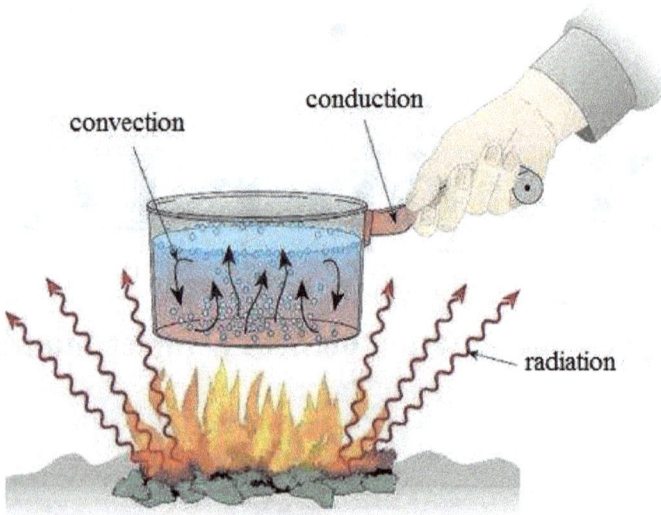

Figure 1. *Diagram of heat transfer by the three modes..*

II.2   Basics:

II.2.1   Temperature:

La température est une grandeur physique mesurée à l'aide d'un thermomètre.

Temperature is a physical quantity measured with a thermometer.

The temperature in the SI is expressed in ° C (degrees Celsius), we meet the temperature in the literature in degrees fahrenheit (° F) and degrees kelvin (° K).

Conversion between different temperature units

°K=°C+ 273

°C=5/9(°F-32) [23]

II.2.2   Heat :

Heat is a form of energy (energy of movement of molecules) that goes from one point hot (higher temperature) to a cold point (lower temperature), The heat is measured in watts (joule / second). [23]

II.2.3   The heat flow:

It is the power exchanged by the surface of the plate, it is measured in watts, it is expressed:

$$\Phi = \frac{1}{A}\frac{dQ}{dt} \qquad\qquad (II.1)$$

Where A is the area of the surface (m^2)[24]

II.2.4   The heat flux density :

As the power exchanged by a unit area of the plate, in watts per square meter. it expresses itself [24]:

$$\varphi = \frac{\Phi}{A} \qquad\qquad (II.2)$$

II.2.5   Temperature gradient:

The temperature gradient is the vector which characterizes at a given point the variation of the temperature function. gradT Defined along the three axes Ox, Oy and Oz by: [24]

$$\overrightarrow{gradT} \begin{cases} \dfrac{\partial T}{\partial x} \\[4pt] \dfrac{\partial T}{\partial x} \\[4pt] \dfrac{\partial T}{\partial z} \end{cases}$$

II.2.6   Power:

Is the ratio of the energy supplied or absorbed over the unit of time. The legal unit is the watt (W). [23]

II.3   Heat transfer modes:

There are three mechanisms of heat exchange between material media:

II.3.1   Convection:

II.3.1.1   Definition:

Convection is a mode of heat transfer when accompanied by mass transfer, this mode most often occurs between a moving fluid and a solid wall. The convection phenomenon is a transfer due to macroscopic movements. [25]

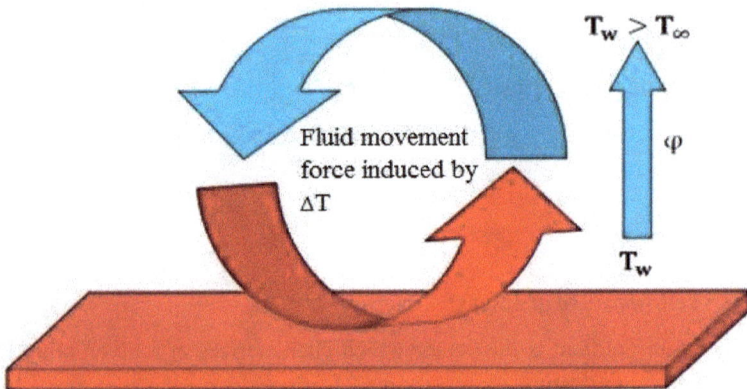

Figure 2. Diagram of convection heat transfer.

Convection is the most important mechanism of heat transfer between a Solid surface and fluid. Two types of convection are generally distinguished:

II.3.1.2   Types of convection :

II.3.1.2.1 Free or natural convection :

Natural convection is caused by an internal force field, in this phenomenon the fluid movement is created by density differences and differences of existing temperatures in the fluid. These applications are: home heating, formation of ocean current... etc. [22]

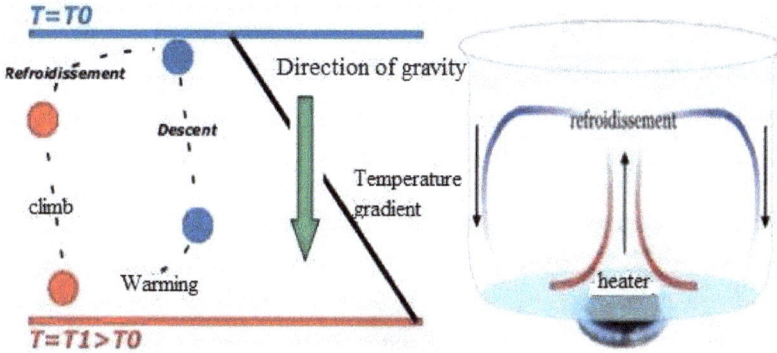

Figure 3. Physical principle of natural convection

II.3.1.2.2 Forced convection:

Forced convection is set in motion by the action of an external force field (pump, fan, etc.). [25]

Example of forced convection: a hair dryer in which the ambient air is blown by a fan through an electric heating resistance.

II.3.1.3   Calculation of the heat exchange coefficient by convection h:

II.3.1.3.1 Newton's law:

Newton's law expresses the mean flux Φ exchanged between a solid and a fluid through their contact surface A [22] :

$$\Phi = hA(\theta_i - \theta_e) \qquad\qquad (II.3)$$

h : Is called the heat exchange coefficient by convection. ($w / m^2 C°$)

($\theta_i - \theta_e$): Difference between the temperature of the wall and the fluid.

In forced convection in the absence of natural convection, the exchange coefficient h by convection is independent of the temperature difference between the wall and the fluid, but it depends on the following six sizes:

U_m : Average fluid velocity

ρ : Volumic masses of the fluid

C_p : Specific heat of the fluid

μ : Dynamic viscosity of the fluid

λ : Thermal conductivity of the fluid

D : Characteristic dimension of the exchange surface

Dimension From these sizes, we define the dimensionless numbers:

II.3.1.3.2 The dimensionless numbers :

- Nusselt number: $\qquad N_U = \dfrac{hD}{\lambda}$

Nu : the Nusselt number characterizes the heat exchange between the fluid and the wall.[26]

- Reynolds number: $\qquad R_e = \dfrac{\rho U_m D}{\mu}$

R_e : The Reynolds number characterizes the flow regime of the fluid

$R_e < 2000$: Laminar flow

$2000 < R_e < 3000$: Intermediate flow

$R_e > 3000$: Turbulent flow

- Number of Prandtl: $\qquad P_r = \dfrac{\mu C_p}{\lambda}$

Pr : the Prandtl number characterizes the thermal properties of the fluid

- The Grashoff number :

$$Gr = \frac{\beta g \rho^2 L^3 \Delta T}{v^2}$$

β : Coefficient of expansion.

g : Acceleration of gravity.

ΔT : Characteristic temperature difference. [26]

The heat exchange coefficient by convection: $\quad h = \dfrac{N_U \lambda}{D}$.

II.3.1.4   Applications:

The applications of convection heat transfer are far too numerous to cite all of them They intervene each time we heat or cool a liquid or a gas, whether it's boiling water in a pan, the central heating radiator, the radiator associated with the engine of a car or the heat exchanger in a process, evaporator or condenser.

Convection applies even if the heat exchange surface is not materialized by a wall, this is the case with condensers by mixing or atmospheric refrigerants, even hot air dryers…[27]

II.3.2   Radiation:

II.3.2.1   Definition :

Radiation is the mechanism by which heat is transmitted from medium to high temperature to another at low temperature when these places media are separated in space.

This transfer mode does not require any material support and can therefore be carried out in the void. In general, the sources of radiation are solids and the radiation is done by the surface.[25]

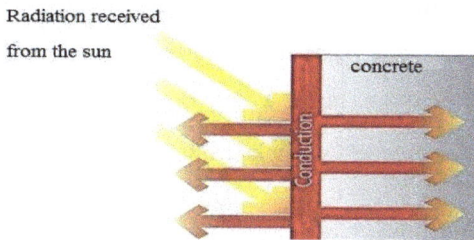

Figure 4. Diagram of Radiant heat transfer scheme

II.3.2.2   Basic law of thermal radiation:

The law **STEFAN-Boltzmann** expresses the radiant energy flux emitted by a ideal surface is proportional to the air of this surface. it is at the fourth power of the absolute temperature T of the surface [22].

$$\Phi = \varepsilon \sigma S T^4 \qquad\qquad (II.4)$$

Or:

Φ : The heat flux in watts.

S : The area of the object in square meters.

ε : Form factor between surfaces i and j.

T : The temperature in Kelvin.

σ: A constant called constant of law of **STEFAN-BOLTZMAN** ($\sigma = 5.67 \times 10^{-8}$ $W / m^2 K^4$)

II.3.2.3 Radiation structure:

Radiation is a mode of energy exchange by emission and absorption of Electromagnetic radiation. The heat exchange by radiation takes place according to the process:

- **Emission:** The energy supplied to the source is converted into energy electromagnetic.
- **Transmission:** The transmission of electromagnetic energy takes place by wave propagation with possible absorption by the medium crossed.
- **Reception:** At reception there is conversion of incident electromagnetic radiation in thermal energy (absorption). [28]

II.3.2.4 Radiation spectrum:

Electromagnetic radiation is made up of waves traveling at the speed of light ($L_\theta = \dfrac{d\Phi}{d\Omega ds . \cos\theta}$ $c = 3 \times 10^8 m s^{-1}$ in the void). We can distinguish them by their wavelength λ or their frequency v linked by: $c = \lambda.v$. We also meet the wave number, linked by: $v = \dfrac{1}{\lambda}$

Visible radiation occupies a narrow band of the spectrum, at wavelengths between 0.38 and 0.78 µm. Shorter wavelengths (higher frequencies high) form ultraviolet radiation, then x and γ. Longer wavelengths than the visible form infrared radiation then microwave. [28]

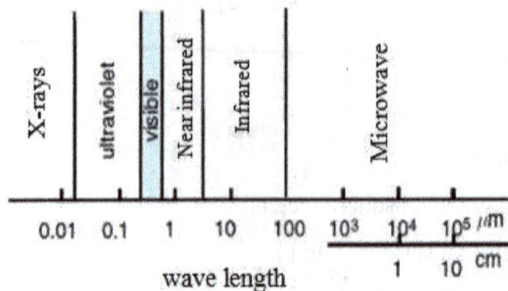

Figure 5. Classification of radiation as a function of wavelength

II.3.2.5   The energy quantities of radiation:

II.3.2.5.1 Energy emitance (Exitance) [M]:

It is the total flux emitted by a surface compared to the unit of this surface. We define [24]:

$$M = \frac{d\Phi}{ds}$$

$$(II.5)$$

How she can be so total

$$M_T = \int_0^\infty M_{\lambda T} d\lambda$$

$$(II.6)$$

II.3.2.5.2 The luminance [L]:

The luminance corresponds to the ratio of the total flux d emitted by a surface dS 'in an angle solid dΩ. The surface dS 'is that the surface dS seen from the direction Ox which makes an angle θ with the normal to this surface. It is expressed by[24] :

$$L_{o.x} = \frac{d\Phi_{o.x}}{d\Omega ds.\cos\theta}$$

$$(II.7)$$

How It can be so total.

$$L_{o.x} = \int_0^\infty L_{o.x}.d\lambda$$

$$(II.8)$$

II.3.2.6   Applications:

Infrared radiation is applied in a large number of industrial processes. Its action on matter is essentially thermal and the main applications concern:

- Drying (paper, cardboard, textiles, etc.);
- Cooking (dyes, primers, coatings, etc.);
- heating (before forming of various materials, heat treatments, welding, heating of workstations ...);
- Polymerizations (inks, coatings, packaging, etc.);
- Sterilization (pharmaceutical bottles, various food products, etc.).

Ultraviolet radiation is used in the field of cross linking of plastic films and polymerizations of organic products such as printing inks, lacquers and varnish, operations which are often improperly called drying. [27]

II.3.3   Conduction:

II.3.3.1   Definition:

Conduction is defined as the mode of heat transfer that is transmitted. energy transfer takes place from one region at high temperature to another at low temperature, inside a solid medium (liquid or gas under certain conditions), or between different backgrounds.[29]

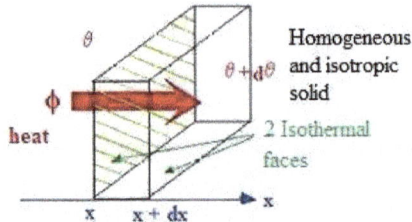

Figure 6. Conduction in an elementary layer of a flat wall

II.3.3.2   FOURIER's law :

The solution of the heat equation by conduction has been exposed by the mathematician French **JBJ FOURIER** in 1822, to fully understand this law, it is necessary to recognize certain number of physical magnitudes. And give by [24]:

$$d^2Q = -k \overrightarrow{.gradT} . \vec{n} \, dA \, dt \qquad\qquad (II.9)$$

II.3.3.2.1 The heat flow:

The heat flow is expressed:

$$\Phi = -k.A.\frac{\Delta\theta}{e} \qquad\qquad (II.10)$$

II.3.3.2.2 Thermal resistance:

According to ohm's law $U = R.I$:

$$U = R.I \qquad \Longleftrightarrow \qquad \Delta\theta = R_t \Phi$$

U : Electric voltage $\Delta\theta$: Temperature deference

R : Electrical resistance. R_t : Thermal resistance.

I : Electric intensity Φ : Heat flux.

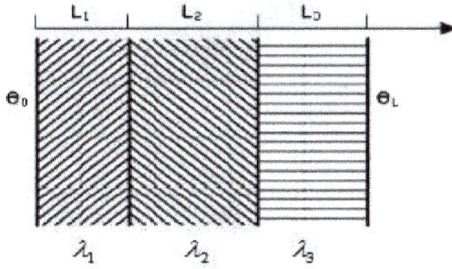

Figure 7. Layout of the multilayer wall

The thermal resistance is:

$$R_t = \frac{e}{\lambda.A} \tag{II.11}$$

R_t : is the thermal resistance.

And the heat flow through the wall is:

$$\Phi = \frac{\theta_0 - \theta_L}{\sum R_t} \tag{II.12}$$

II.3.3.3   Applications:

Conductive heat transfer characterizes all heat transfers that take place in the walls separating two bodies at different temperatures. It's the case with exchange surfaces of heat exchangers, but it is also that of walls and glazing of a building of tanks containing hot or cold liquids, walls of furnaces, etc ... [27]

II.4   Conclusion:

After this exposure, we come to conclude that:

The three mechanisms of heat transfer are: convection is a transfer by transport, radiation is a remote transfer all the more important as the temperature is high and recently the conduction is a transfer in the mass.

Chapter III:

General information on the resolution methods

III.   Chapter III. General information on the resolution methods

III.1 Introduction :

The numerical resolution is based on the method of discretization of the equations of the problem.

This method consists in transforming a differential equation into an algebraic equation, easy to solve (discretized equation). The most common discretization methods used are:

- Finite differences.

- Finite elements.

- Finished volumes.

These numerical methods are very useful for the resolution of the mechanical problems of fluids, heat transfer, material transfer...

III.2 The three main families of methods:

To go from a continuous exact problem governed by a PDE to the discrete approximate problem, it is there are three main families of methods:

III.2.1   Finite elements (FE):

III.2.1.1 Definition:

The term "finite element" is used for the first time by GLOUGH, and from then on, there is a rapid development of the method between 1965 and 1975.

The finite element method is a method of approximation of the solutions of equations with partial derivatives which is built starting from an equivalent formulation of the problem to solve, which can be written schematically in the following form: $[A].\{U\}=\{L\}$

Or

$\{U\}$: is the vector of the unknowns,

$[A]$: is a matrix,

$\{L\}$: is a vector.

The method allows to process complex geometries unlike finite differences as well as many theoretical results on convergence, on the other hand it requires a high cost of computation time and memory.

Many structural calculation codes are based on Finite Elements: ANSYS, CADDS, CATIA...[30]

III.2.1.2 General principles:

The finite element method has principles such as:

- Identify subdomains Ω_e geometrically simple which pave the domain;
- Define an approximate function on each subdomain.

We can therefore imagine a certain number of characteristics of this construction:

- Paving of the field Ω_e by the subdomains Ω_e should be as precise as possible.
- The approximate function on the subdomain must respect conditions of continuity between the different subdomains.
- The function approximated on the subdomain must have coherent properties with the differentiability conditions and related to the physical description of the solution (this which may involve using a weakened formulation for example) [27]

III.2.2   Finite element mesh:

Consists of dividing the geometry into a finite number of the domain (elementary domain). There are different types of elements:

- Linear element (1D).
- Surface element (2D).
- Voluminal element (3D).

For structures in (2D) the mesh elements are often triangles but it can also be in other form, the possible forms in this case are the following [31]:

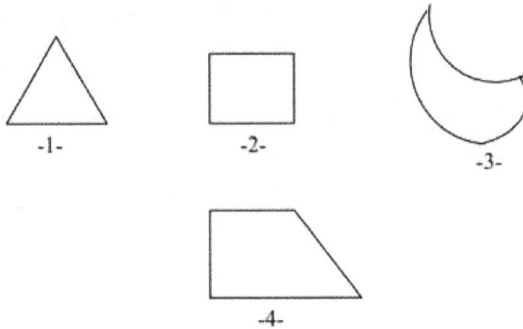

Figure 8. the possible shapes of elementary elements in (2D)

III.2.3   Finite differences (FD):

III.2.3.1 Definition:

The method consists in replacing the partial derivatives by divided differences or combinations of point values of the function in a finite number of discrete points or mesh nodes.

This method as it has advantages, has disadvantages:

The advantages: great ease of writing and low cost of calculation.

The disadvantages: limitation to simple geometries, difficulties in taking into account boundary conditions of the Neumann type. [30]

III.2.3.2   Principle:

Elle est basée sur le développement de Taylor de la fonction inconnue, autour d'un point x_0 et pour une fonction $f(x)$ continue et n fois dérivable, nous pouvons écrire :

It is based on the Taylor expansion of the unknown function, around a point x_0 and for a function $f(x)$ continuous and n times differentiable, we can write:

$$\frac{df(x_0)}{dx} = \frac{f(x_0 + \Delta x) + f(x_0 - \Delta x) - 2f(x_0)}{2.\Delta x} \qquad (III.1)$$

$$\frac{d^2 f(x_0)}{dx^2} = \frac{f(x_0 + \Delta x) + f(x_0 - \Delta x) - 2f(x_0)}{\Delta x^2} \qquad (III.2)$$

This method consists in having a simple geometry and a single homogeneous medium. [31]

III.2.3.3 Limited development:

According to **Taylor**:

$$f(x+h) = \frac{1}{0!}f^{\circ}(x) + \frac{h^1}{1!}f'(x) + \frac{h^2}{2}f'(x) + ... \frac{h^n}{n!}f''(x)$$
(III .3)

We apply the expansion of order 1 we find:

$$f(x+h) = f(x) + hf'(x) + 0(h)$$
(III .4)

We have Taylor's theorems for the points:

- 3 points :

Centered diagram: $f'(x) = \frac{f(x+h) - f(x-h)}{2h} - 0(h^2)$

Diagram before: $f'(x) = \frac{f(x+1) - f(x)}{h} + 0(h)$

Back diagram: $f'(x) = \frac{f(x) - f(x-1)}{h} + 0(h)$

III.2.3.4 Solving the heat equation by finite differance:

III.2.3.4.1 Explicit method

We consider the one-dimensional heat equation:

$$\frac{\partial T}{\partial t} = \alpha \frac{\partial^2 T}{\partial x^2} \quad \text{à} \quad \begin{array}{l} 0 \le x \le L \\ 0 \le x \le 1 \\ 0 \le x \le 0.1 \end{array} \quad , t > 0$$

With the following boundary and initial conditions

$$\begin{cases} T(0,t) = 0 \\ T(L,t) = 0 \\ T(x,0) = \sin(\pi x) + \sin(3\pi x) \end{cases}$$

We calculated the unknown temperatures on the field [0.01] by the explicit diagram.

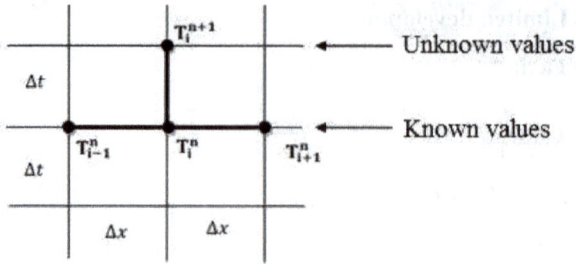

Figure 9. the nodes of the finite differences of the simple explicit diagram

$$r = \frac{\Delta t}{\Delta x^2} \ ,$$

$$T_i^{j+1} = rT_{i-1}^j - (1-2r)T_i^j + rT_{i+1}^j \tag{III.5}$$

$$j = 1,2,3... \qquad i = 1,2,...,n-1$$

The calculation procedure is as follows:

- Start the calculation with $j=0$. Calculate T_i^1, i=1, 2,…, n-1 by the equation, as long as the temperatures on the right side of the equation are defined by the boundary conditions.

- Meter in the second step $n=1$ and calculate $T_i^2, i=1,2,...,n-1$ using the temperatures calculated in the previous step.

- Repeat the procedure for each time step and continue the calculations until one instant when a temperature value is reached.

III.2.3.4.2 Implicit method:

Let the 1D heat equation is:

$$\frac{\partial T}{\partial t} = \frac{\partial^2 T}{\partial x^2}$$

Défini sur le domaine suivant :

$$\begin{cases} x \in [0,3] \\ t \in [0,1] \end{cases}$$

With the boundary conditions:

$$\begin{cases} T(0,t) = 300 \\ T(3,t) = 400 \end{cases}$$

And initial condition:

$T(x,0) = 100$

The unknown temperatures were calculated by the implicit scheme.

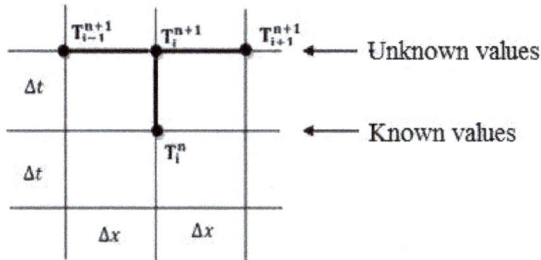

Figure 10. the nodes of the finite differences of the simple implicit diagram

$$r = \alpha \frac{\Delta t}{\Delta x^2}$$

$$T_i^{j+1} = \frac{w}{(1+2r)}(T_i^j + rT_{i-1}^{j+1} + rT_{i+1}^{j+1}) + T_i^{j+1}(1+w) \qquad (III.6)$$

III.2.4   Finite volumes (FV):

III.2.4.1   Definition:

For the first time in 1971, the finite volume method was described by **PATANKAR** and **SPALDING** and published in 1980 by **PATANKAR**, this method is a method of discretization one uses to numerically solve the differential equations with partial derivatives, as well as the phenomena studied by physicists and engineers, the fluid mechanics, mass and heat transport. , Many numerical simulation codes in fluid mechanics are based on this method: Fluent, StarCD, CFX, Fine Turbo, elsA ...

The finite volume method is suitable for the equations of fluid mechanics

- Mass conservation equation,
- Conservation equation of momentum,
- Energy conservation equation. [30]

The method makes it possible to process complex geometries with shaped volumes arbitrary, as well as to determine more natural boundary conditions of type Neumann. On the other hand, few theoretical results of convergence. [27]

III.2.4.2 Principle

The principle of the finite volume method is divided into a finite number of subdomains elementary, called control volume, as shown in figure (11). Each of these last includes a node called principal node P, the points E and W (East and West) are neighbors in the x direction, while N and S (North and South) are those in the y direction. The control volume surrounding P is shown by the broken lines. The faces of the volume of control are located at points e and w in the x direction, n and s in the y direction [32].

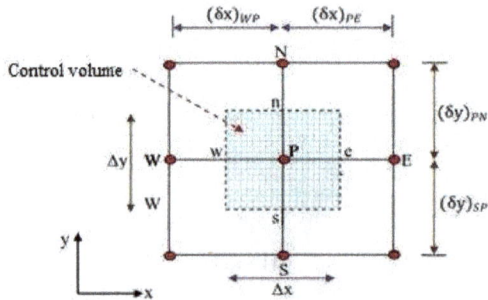

Figure 11. Typical control volume for a 2D situation

III.2.4.3   Concept of mesh:

III.2.4.3.1 Definition:

It is the subdivision of the field of study into longitudinal and transverse grids, of which the intersection represents a node, where we find the central point P and the components u and v of velocity vector, which are in the middle of the segments.

On the other hand, we carried out a research on another definition, the mesh is: The field of study is divided into a number of control volumes, each point of the domain is located using indices (i, j). Figure (12)

The scalar quantities are stored in the node P of the mesh; the vector quantities u and v are stored in the middle of the segments connecting the nodes.

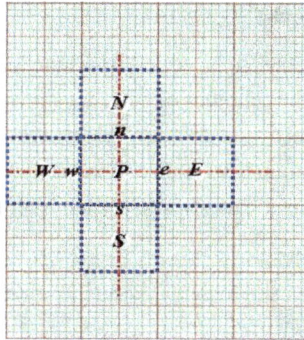

Figure 12. Two-dimensional control volume

The control volume of the transverse component u is shifted to the right with respect to the main control volume see Figure (13), that of the longitudinal component v is shifted up (14). This type of mesh, known as "shifted mesh" [32]

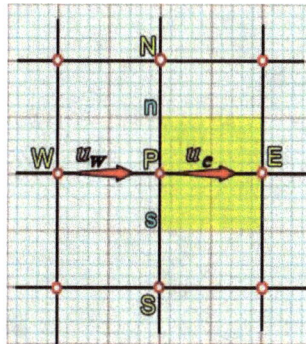

Figure 13. Control volume shifted to the right

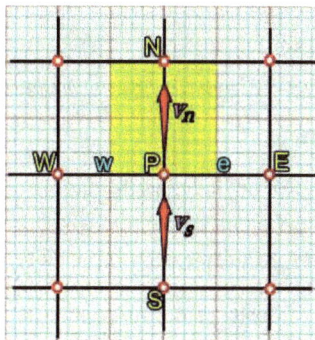

Figure 14. Control volume shifted up

III.2.4.3.2 Choice of mesh:

After creating the geometry, it is very important to choose a suitable mesh to process the problem, The choice of the mesh is an essential point in the precision and the correctness of numerical results. To do this, we must determine the optimal parameters and choose a networking strategy that meets our objectives, Among these parameters, we can mention:

- The number of meshes,
- The distance between the meshes (concentration of the meshes),
- The shape of the mesh,
- The deformation parameters for the case of the deformable mesh. [33]

We distinguish several types of meshes:

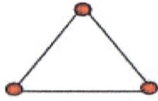

2D element of "triangle" type

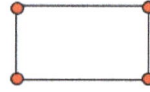

2D element of "quadrilateral" type

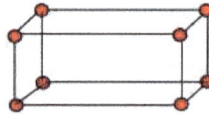

3D element of the "hexahedron" type.

Figure 15. Types of meshes

III.2.4.3.3 Example 1D:

Problème de diffusion 1D de la chaleur dans une barre "AB" horizontal de longueur L avec un coefficient de conductivité k et une section A avec des conditions aux limites et T_B sans terme source.

1D diffusion problem of heat in a horizontal bar "AB" of length L with a conductivity coefficient k and a section A with boundary conditions T_A and T_B without source term.

- **Solution:**

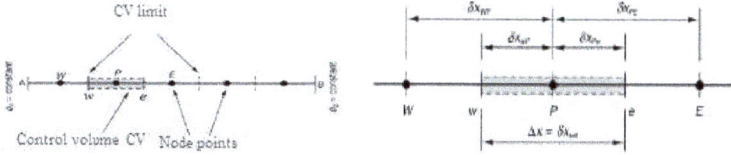

Figure 16. the volume domain of discrete controls

- **Discretization:**

$$\int_{cv}\underbrace{\frac{\partial(\rho T)}{\partial t}}_{\substack{= \\ 0}}dv + \int_{cv}\underbrace{div\,(\rho T u)}_{\substack{= \\ 0}}dv = \int_{cv}\underbrace{div\,(k\,\overline{gradT})}_{\substack{term \\ diffusive}}dv + \int_{cv}\underbrace{ST}_{\substack{term \\ source}}dv \qquad (III.7)$$

$$\int_{v} div\,(k\,\overline{gradT})dv + \int_{v} ST\,dv = 0 \qquad (III.8)$$

$$\int_{v}\frac{\partial}{\partial x}(k\,\overline{gradT})dA\,dx + \int_{v} STdv = 0 \qquad (III.9)$$

$$(k\,\frac{dT}{dx})_{w}^{e}\,A + Su = 0 \qquad (III.10)$$

$$(\underbrace{\frac{k_{e}A_{e}}{\Delta x} + \frac{k_{w}A_{w}}{\Delta x} - S_{P}}_{a_{P}})T_{P} = \underbrace{\frac{k_{e}A_{e}}{\Delta x}}_{a_{E}}T_{E} + \underbrace{\frac{k_{w}A_{w}}{\Delta x}}_{a_{W}}T_{W} + Su \qquad (III.11)$$

$$a_{P}T_{P} = a_{W}T_{W} + a_{E}T_{E} + Su \qquad (III.12)$$

$$a_{P} = a_{W} + a_{E} - S_{p} \qquad (III.13)$$

After the discretization we found:

- **nodes (02, 03,04) :**

$$(\underbrace{\frac{k_{e}A_{e}}{\Delta x} + \frac{k_{w}A_{w}}{\Delta x}}_{a_{P}})T_{P} = \underbrace{\frac{k_{e}A_{e}}{\Delta x}}_{a_{E}}T_{E} + \underbrace{\frac{k_{w}A_{w}}{\Delta x}}_{a_{W}}T_{W} \qquad (III.14)$$

- **nodes (01) :**

$$(\underbrace{\frac{k_{e}A_{e}}{\Delta x} + 0 + \frac{2k_{w}A_{w}}{\Delta x}}_{a_{P}})T_{P} = \underbrace{\frac{k_{e}A_{e}}{\Delta x}}_{a_{E}}T_{E} + T_{W}(0) + \underbrace{\frac{2k_{w}A_{w}}{\Delta x}T_{A}}_{Su} \qquad (III.15)$$

- **nodes (** **05) :**

$$\underbrace{(\frac{k_w A_w}{\Delta x}+0+\frac{2k_e A_e}{\Delta x})}_{a_P}T_P = \underbrace{\frac{k_w A_w}{\Delta x}}_{a_E}T_W + \underbrace{T_E(0)}_{q_r} + \underbrace{\frac{2k_e A_e}{\Delta x}T_B}_{Su} \qquad\qquad (III.16)$$

The following table deals with the results of five nodes

N	a_P	a_W	a_E	Su	Sp
1	$\dfrac{k_e A_e}{\Delta x}+\dfrac{2k_w A_w}{\Delta x}$	0	$\dfrac{k_e A_e}{\Delta x}$	$\dfrac{2k_w A_w}{\Delta x}T_A$	$\dfrac{2k_w A_w}{\Delta x}$
2	$\dfrac{k_w A_w}{\Delta x}+\dfrac{k_e A_e}{\Delta x}$	$\dfrac{k_w A_w}{\Delta x}$	$\dfrac{k_e A_e}{\Delta x}$	0	0
3	$\dfrac{k_w A_w}{\Delta x}+\dfrac{k_e A_e}{\Delta x}$	$\dfrac{k_w A_w}{\Delta x}$	$\dfrac{k_e A_e}{\Delta x}$	0	0
4	$\dfrac{k_w A_w}{\Delta x}+\dfrac{k_e A_e}{\Delta x}$	$\dfrac{k_w A_w}{\Delta x}$	$\dfrac{k_e A_e}{\Delta x}$	0	0
5	$\dfrac{k_w A_w}{\Delta x}+\dfrac{2k_e A_e}{\Delta x}$	$\dfrac{k_w A_w}{\Delta x}$	0	$\dfrac{k_e A_e}{\Delta x}T_B$	$\dfrac{2k_e A_e}{\Delta x}$

Tableau1. The results of five nodes

III.3 Conclusion :

Throughout this chapter, we have briefly presented the three methods of resolutions, examples on each method. In our study we used the finite volume method to solve the heat equation. This equation is discretized in the next chapter.

Chapter IV:

Governing equation model

IV.   Chapter IV. Governing equation model

IV.1 Introduction:

This chapter is a passage on the modeling of the heat equation in a plate. First, the general heat equation will be presented and the method of resolution of this equation by the finite volume method which will be the main subject of this this chapter.

IV.2 Modeling of heat equation:

Numerical modeling is at the heart of applied sciences and plays a fundamental role in almost all science and engineering disciplines. Modeling or digital simulation, consists in representing a physical phenomenon by a mathematical model under form of very large systems of equations which are solved using programs.

IV.3 Discretization:

Before discretizing the heat equation presented in the previous chapter, we will specify the boundary conditions.

IV.3.1   Governing equation:

The heat equation for a 2D problem is as follows:

$$\underbrace{\frac{\partial(\rho T)}{dt}}_{\substack{term \\ temp}} + \underbrace{div\,(\rho T u)dv}_{\substack{term \\ convective}} = \underbrace{div\,(k\,\overline{gradT})}_{\substack{term \\ diffusive}} + \underbrace{ST}_{\substack{term \\ source}} \qquad\qquad (IV\ .1)$$

We consider a rectangular plate of length L and height H:

IV.3.1.1   The first case:

IV.3.1.1.1 Boundary conditions of the "imposed temperature" type (Dirichlet)

It is assumed that the area of the border $x=0$ is maintained at uniform temperature T_1 the boundary surface at $x=L$ the uniform temperature is maintained T_3, and the surface of the border $y=0$ is maintained at temperature T_4 and the surface of the border $y=H$ is maintained at uniform temperature T_2.

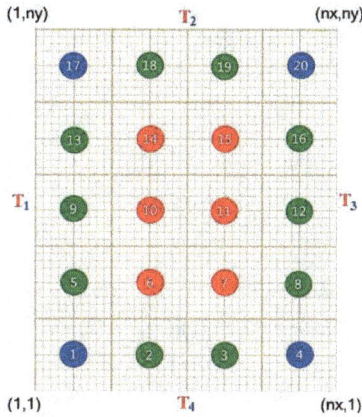

*Figure 17. CL **Dirichlet** Control Volume*

IV.3.1.1.2 Numerical resolution of the heat equation in 2D by the method of finished volumes:

$$\int_v \frac{\partial(\rho T)}{dt}dv + \int_v div\,(\rho T u)dv = \int_v div\,(k\,\overline{gradT}\,)dv + \int_v ST\,dv$$

$$\underbrace{}_{\substack{=\\0}} \qquad \underbrace{}_{\substack{=\\0}} \qquad \underbrace{}_{\substack{term\\diffusive}} \qquad \underbrace{}_{\substack{term\\source}}$$

$$\int_v div\,(k\,\overline{gradT}\,)dv = 0$$

$$\int_v \frac{\partial}{\partial x}\left(k\,\frac{\partial T}{\partial x}\right)dA_x\,dx + \int_v \frac{\partial}{\partial y}\left(k\,\frac{\partial T}{\partial y}\right)dA_y\,dy$$

$$\int_s (k\,\frac{dT}{dx})dA_x + \int_s (k\,\frac{dT}{dy})dA_y = 0 \qquad (VI.1)$$

$$k_e A_e \left.\frac{\partial T}{\partial x}\right|_e - k_w A_w \left.\frac{\partial T}{\partial x}\right|_w + k_s A_s \left.\frac{\partial T}{\partial y}\right|_s - k_n A_n \left.\frac{\partial T}{\partial y}\right|_n = 0 \qquad (VI.2)$$

$$\frac{k_e A_e}{\Delta x}(T_E - T_P) - \frac{k_w A_w}{\Delta x}(T_P - T_W) + \frac{k_n A_n}{\Delta y}(T_N - T_P) - \frac{k_s A_s}{\Delta y}(T_P - T_S) = 0 \qquad (VI.3)$$

$$a_P T_P = a_W T_W + a_E T_E + a_S T_S + a_N T_N + Su \qquad (VI.4)$$

$$a_P = a_W + a_E + a_S + a_N - S_p \qquad (VI.5)$$

- **Node (1)**

$$\frac{k_e A_e}{\Delta x}(T_E - T_P) - \frac{2k_w A_w}{\Delta x}(T_P - T_1) + \frac{k_n A_n}{\Delta y}(T_N - T_P) - \frac{2k_s A_s}{\Delta y}(T_P - T_4) = 0$$

$$(\underbrace{\underbrace{\frac{k_e A_e}{\Delta x}}_{a_E} + \underbrace{\frac{k_n A_n}{\Delta y}}_{a_N} + \underbrace{\frac{2k_w A_w}{\Delta x} + \frac{2k_s A_s}{\Delta y}}_{-S_P}}_{a_P})T_P = \frac{k_e A_e}{\Delta x}T_E + \frac{k_n A_n}{\Delta y}T_N + \underbrace{\frac{k_w A_w}{\Delta x}T_1 + \frac{2k_s A_s}{\Delta y}T_4}_{S_u}$$

- **Node (4)**

$$\frac{2k_e A_e}{\Delta x}(T_3 - T_P) - \frac{k_w A_w}{\Delta x}(T_P - T_W) + \frac{k_n A_n}{\Delta y}(T_N - T_P) - \frac{2k_s A_s}{\Delta y}(T_P - T_4) = 0$$

$$(\underbrace{\underbrace{\frac{k_w A_w}{\Delta x}}_{a_W} + \underbrace{\frac{k_n A_n}{\Delta y}}_{a_N} + \underbrace{\frac{2k_e A_e}{\Delta x} + \frac{2k_s A_s}{\Delta y}}_{-S_P}}_{a_P})T_P = \frac{k_w A_w}{\Delta x}T_W + \frac{k_n A_n}{\Delta y}T_N + \underbrace{\frac{k_e A_e}{\Delta x}T_3 + \frac{k_s A_s}{\Delta y}T_4}_{S_u}$$

- **Node (17)**

$$\frac{k_e A_e}{\Delta x}(T_E - T_P) - \frac{2k_w A_w}{\Delta x}(T_P - T_1) + \underbrace{\frac{2k_n A_n}{\Delta y}(T_2 - T_P)}_{0} - \frac{k_s A_s}{\Delta y}(T_P - T_S) = 0$$

$$(\underbrace{\underbrace{\frac{k_e A_e}{\Delta x}}_{a_E} + \underbrace{\frac{k_s A_s}{\Delta y}}_{a_S} + \underbrace{\frac{2k_n A_n}{\Delta y} + \frac{2k_w A_w}{\Delta x}}_{-S_P}}_{a_P})T_P = \frac{k_e A_e}{\Delta x}T_E + \frac{k_s A_s}{\Delta y}T_S + \underbrace{\frac{2k_w A_w}{\Delta x}T_1 + \frac{2k_n A_n}{\Delta y}T_2}_{S_u}$$

- **Node (20)**

$$\frac{2k_e A_e}{\Delta x}(T_3 - T_P) - \frac{k_w A_w}{\Delta x}(T_P - T_w) + \frac{2k_n A_n}{\Delta y}(T_2 - T_P) - \frac{k_s A_s}{\Delta y}(T_P - T_S) = 0$$

$$(\underbrace{\underbrace{\frac{k_w A_w}{\Delta x}}_{a_W} + \underbrace{\frac{k_s A_s}{\Delta y}}_{a_S} + \underbrace{\frac{2k_e A_e}{\Delta x} + \frac{2k_n A_n}{\Delta y}}_{-S_P}}_{a_P})T_P = \frac{k_w A_w}{\Delta x}T_W + \frac{k_s A_s}{\Delta y}T_S + \underbrace{\frac{2k_e A_e}{\Delta x}T_3 + \frac{2k_n A_n}{\Delta y}T_2}_{S_u}$$

- **Nodes (05, 09,13)**

$$\frac{k_e A_e}{\Delta x}(T_E - T_P) - \frac{2k_w A_w}{\Delta x}(T_P - T_1) + \frac{k_n A_n}{\Delta y}(T_N - T_P) - \frac{k_s A_s}{\Delta y}(T_P - T_S) = 0$$

$$(\underbrace{\underbrace{\frac{k_e A_e}{\Delta x}}_{a_E} + \underbrace{\frac{k_s A_s}{\Delta y}}_{a_S} + \underbrace{\frac{k_n A_n}{\Delta y}}_{a_N} + \underbrace{\frac{2k_w A_w}{\Delta x}}_{-S_P}}_{a_P})T_P = \frac{k_e A_e}{\Delta x}T_E + \frac{k_s A_s}{\Delta y}T_S + \frac{k_n A_n}{\Delta y}T_N + \underbrace{\frac{2k_w A_w}{\Delta x}T_1}_{S_u}$$

- **Nodes (8, 12,16)**

$$\frac{2k_eA_e}{\Delta x}(T_3-T_P)-\frac{k_wA_w}{\Delta x}(T_P-T_W)+\frac{k_nA_n}{\Delta y}(T_N-T_P)-\frac{k_sA_s}{\Delta y}(T_P-T_S)=0$$

$$\underbrace{(\underbrace{\frac{k_wA_w}{\Delta x}}_{a_W}+\underbrace{\frac{k_sA_s}{\Delta y}}_{a_S}+\underbrace{\frac{k_nA_n}{\Delta y}}_{a_N}+\underbrace{\frac{2k_eA_e}{\Delta x}}_{-S_P})}_{a_P}T_P=\frac{k_wA_w}{\Delta x}T_W+\frac{k_sA_s}{\Delta y}T_S+\frac{k_nA_n}{\Delta y}T_N+\underbrace{\frac{2k_eA_e}{\Delta x}}_{Su}T_3$$

- **Nodes (2,3)**

$$\frac{k_eA_e}{\Delta x}(T_E-T_P)-\frac{k_wA_w}{\Delta x}(T_P-T_W)+\frac{k_nA_n}{\Delta y}(T_N-T_P)-\underbrace{\frac{2k_sA_s}{\Delta y}(T_P-T_4)}_{0}=0$$

$$\underbrace{(\underbrace{\frac{k_eA_e}{\Delta x}}_{a_E}+\underbrace{\frac{k_wA_w}{\Delta x}}_{a_W}+\underbrace{\frac{k_nA_n}{\Delta y}}_{a_n}+\underbrace{\frac{2k_sA_s}{\Delta y}}_{-Sp})}_{a_P}T_P=\frac{k_eA_e}{\Delta x}T_E+\frac{k_wA_w}{\Delta x}T_w+\frac{k_nA_n}{\Delta y}T_N+\underbrace{\frac{2k_sA_s}{\Delta y}}_{Su}T_4$$

- **Nodes (18,19)**

$$\frac{k_eA_e}{\Delta x}(T_E-T_P)-\frac{k_wA_w}{\Delta x}(T_P-T_W)+\frac{2k_nA_n}{\Delta y}(T_2-T_P)-\frac{k_sA_s}{\Delta y}(T_P-T_S)=0$$

$$\underbrace{(\underbrace{\frac{k_eA_e}{\Delta x}}_{a_E}+\underbrace{\frac{k_wA_w}{\Delta x}}_{a_W}+\underbrace{\frac{k_sA_s}{\Delta y}}_{a_S}+\underbrace{\frac{2k_nA_n}{\Delta y}}_{-S_P})}_{a_P}T_P=\frac{k_eA_e}{\Delta x}T_E+\frac{k_wA_w}{\Delta x}T_w+\frac{k_sA_s}{\Delta y}T_S+\frac{k_nA_n}{\Delta y}T_2$$

- **Nodes (6,7 ,10 ,11 ,14 ,15)**

$$\frac{k_eA_e}{\Delta x}(T_E-T_P)-\frac{k_wA_w}{\Delta x}(T_P-T_W)+\frac{k_nA_n}{\Delta y}(T_N-T_P)-\frac{k_sA_s}{\Delta y}(T_P-T_S)=0$$

$$\underbrace{(\underbrace{\frac{k_wA_w}{\Delta x}}_{a_W}+\underbrace{\frac{k_eA_e}{\Delta x}}_{a_E}+\underbrace{\frac{k_sA_s}{\Delta y}}_{a_S}+\underbrace{\frac{k_nA_n}{\Delta y}}_{a_N})}_{a_P}T_P=\frac{k_wA_w}{\Delta x}T_W+\frac{k_eA_e}{\Delta x}T_E+\frac{k_sA_s}{\Delta y}T_S+\frac{k_nA_n}{\Delta y}T_N$$

IV.3.1.2   The second case :

IV.3.1.2.1 Boundary conditions of the "zero flux" type (Neumann), "temperature imposed "(Dirichlet)

- Neumann is an "imposed flow" in our case "null imposed flow" (adiabatic)

It is assumed that the area of the border $x = 0$ is maintained at the uniform temperature T_1 and the boundary surface at $y = H$ the uniform temperature is maintained at the uniform temperature T_2, the surface of the border $x = L$ and the boundary surface at $y = 0$ are maintained at the **Neumann** condition

$$\left(\frac{dT}{dy} = 0, \frac{dT}{dx} = 0 \right).$$

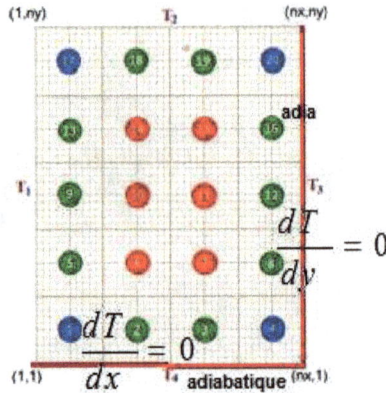

*Figure 18. CL **Neumann** and **Dirichlet** control volume*

IV.3.1.2.2 Numerical resolution of the heat equation in 2D by the method of finished volumes:

$$\underbrace{\int_v \frac{\partial(\rho T)}{dt} dv}_{\substack{= \\ 0}} + \underbrace{\int_v div\,(\rho T u)\,dv}_{\substack{= \\ 0}} = \underbrace{\int_v div\,(k\,\overline{gradT}\,)dv}_{\substack{term \\ diffusive}} + \underbrace{\int_v ST\,dv}_{\substack{term \\ source}}$$

$$\int_v div\,(k\,\overline{gradT}\,)dv = 0$$

$$\int_v \frac{\partial}{\partial x}\left(k\,\frac{\partial T}{\partial x} \right) dA_x\,dx + \int_v \frac{\partial}{\partial y}\left(k\,\frac{\partial T}{\partial y} \right) dA_y\,dy$$

$$\int_s (k\,\frac{dT}{dx})dA_x + \int_s (k\,\frac{dT}{dy})dA_y = 0$$

$$k_e A_e \frac{\partial T}{\partial x}\bigg|_e - k_w A_w \frac{\partial T}{\partial x}\bigg|_w + k_s A_s \frac{\partial T}{\partial y}\bigg|_s - k_n A_n \frac{\partial T}{\partial y}\bigg|_n = 0$$

$$\frac{k_e A_e}{\Delta x}(T_E - T_P) - \frac{k_w A_w}{\Delta x}(T_P - T_W) + \frac{k_s A_s}{\Delta y}(T_P - T_S) - \frac{k_n A_n}{\Delta y}(T_N - T_P) = 0$$

$$a_P T_P = a_W T_W + a_E T_E + a_S T_S + a_N T_N + Su$$

$$a_P = a_W + a_E + a_S + a_N - S_p$$

- **Node (01) :**

$$\frac{k_e A_e}{\Delta x}(T_E - T_P) - \frac{2k_w A_w}{\Delta x}(T_P - T_1) + \frac{k_n A_n}{\Delta y}(T_N - T_P) - \underbrace{\frac{2k_s A_s}{\Delta y}(T_P - T_4)}_{0} = 0 \quad \Leftrightarrow T_P = T_4$$

$$(\underbrace{\frac{k_e A_e}{\Delta x}}_{a_E} + \underbrace{\frac{k_n A_n}{\Delta y}}_{a_N} + \underbrace{\frac{2k_w A_w}{\Delta x}}_{-S_P})T_P = \frac{k_e A_e}{\Delta x}T_E + \frac{k_n A_n}{\Delta y}T_N + \underbrace{\frac{2k_w A_w}{\Delta x}T_1}_{Su}$$

$$\underbrace{}_{a_P}$$

- **Node (04) :**

$$\frac{2k_e A_e}{\Delta x}(T_2 - T_P) - \frac{k_w A_w}{\Delta x}(T_P - T_W) + \frac{k_n A_n}{\Delta y}(T_N - T_P) - \underbrace{\frac{2k_s A_s}{\Delta y}(T_P - T_4)}_{0} = 0 \quad \Leftrightarrow T_P = T_4$$

$$(\underbrace{\frac{k_w A_w}{\Delta x}}_{a_W} + \underbrace{\frac{k_n A_n}{\Delta y}}_{a_N})T_P = \frac{k_w A_w}{\Delta x}T_W + \frac{k_n A_n}{\Delta y}T_N$$

$$\underbrace{}_{a_P}$$

$$Su = 0 \qquad Sp = 0$$

- **Node (17) :**

$$\frac{k_e A_e}{\Delta x}(T_E - T_P) - \frac{2k_w A_w}{\Delta x}(T_P - T_1) + \underbrace{\frac{2k_n A_n}{\Delta y}(T_2 - T_P)}_{0} - \frac{k_s A_s}{\Delta y}(T_P - T_S) = 0$$

$$(\underbrace{\frac{k_e A_e}{\Delta x}}_{a_E} + \underbrace{\frac{k_s A_s}{\Delta y}}_{a_S} + \underbrace{\frac{2k_w A_w}{\Delta x} + \frac{2k_n A_n}{\Delta y}}_{-S_P})T_P = \frac{k_e A_e}{\Delta x}T_E + \frac{k_s A_s}{\Delta y}T_S + \underbrace{\frac{2k_w A_w}{\Delta x}T_1 + \frac{2k_n A_n}{\Delta y}T_2}_{Su}$$

$$\underbrace{}_{a_P}$$

- **Node (20) :**

$$\underbrace{\frac{2k_e A_e}{\Delta x}(T_3 - T_P)}_{0} - \frac{k_w A_w}{\Delta x}(T_P - T_w) + \frac{2k_n A_n}{\Delta y}(T_2 - T_P) - \frac{k_s A_s}{\Delta y}(T_P - T_S) = 0 \quad \Leftrightarrow T_3 = T_P$$

$$(\underbrace{\frac{k_w A_w}{\Delta x}}_{a_W} + \underbrace{\frac{k_s A_s}{\Delta y}}_{a_S} + \underbrace{\frac{2k_n A_n}{\Delta y}}_{-S_P})T_P = \frac{k_w A_w}{\Delta x}T_w + \frac{k_s A_s}{\Delta y}T_S + \underbrace{\frac{2k_n A_n}{\Delta y}T_2}_{Su}$$

$$\underbrace{}_{a_P}$$

- **Nodes (09, 05,13) :**

$$\frac{k_e A_e}{\Delta x}(T_E - T_P) - \frac{2k_w A_w}{\Delta x}(T_P - T_1) + \frac{k_n A_n}{\Delta y}(T_N - T_P) - \frac{k_s A_s}{\Delta y}(T_P - T_S) = 0$$

$$(\underbrace{\frac{k_e A_e}{\Delta x}}_{a_E} + \underbrace{\frac{k_s A_s}{\Delta y}}_{a_S} + \underbrace{\frac{k_n A_n}{\Delta y}}_{a_N} + \underbrace{\frac{2k_w A_w}{\Delta x}}_{-S_P})T_P = \frac{k_e A_e}{\Delta x}T_E + \frac{k_s A_s}{\Delta y}T_S + \frac{k_n A_n}{\Delta y}T_N + \underbrace{\frac{2k_w A_w}{\Delta x}T_1}_{Su}$$

$$\underbrace{\frac{}{}}_{a_P}$$

- ### Nodes (08, 12,16) :

$$\underbrace{\frac{2k_e A_e}{\Delta x}(T_3 - T_P)}_{0} - \frac{k_w A_w}{\Delta x}(T_P - T_W) + \frac{k_n A_n}{\Delta y}(T_N - T_P) - \frac{k_s A_s}{\Delta y}(T_P - T_S) = 0 \quad \Leftrightarrow T_3 = T_P$$

$$(\underbrace{\frac{k_w A_w}{\Delta x}}_{a_W} + \underbrace{\frac{k_s A_s}{\Delta y}}_{a_S} + \underbrace{\frac{k_n A_n}{\Delta y}}_{a_N})T_P = \frac{k_w A_w}{\Delta x}T_W + \frac{k_s A_s}{\Delta y}T_S + \frac{k_n A_n}{\Delta y}T_N$$

$$\underbrace{\frac{}{}}_{a_P}$$

$$Su = 0 \qquad Sp = 0$$

- ### Nodes (02,03) :

$$\frac{k_e A_e}{\Delta x}(T_E - T_P) - \frac{k_w A_w}{\Delta x}(T_P - T_W) + \frac{k_n A_n}{\Delta y}(T_N - T_P) - \underbrace{\frac{2k_s A_s}{\Delta y}(T_P - T_4)}_{0} = 0 \quad \Leftrightarrow T_P = T_B$$

$$(\underbrace{\frac{k_e A_e}{\Delta x}}_{aE} + \underbrace{\frac{k_w A_w}{\Delta x}}_{a_W} + \underbrace{\frac{k_n A_n}{\Delta y}}_{a_N})T_P = \frac{k_e A_e}{\Delta x}T_E + \frac{k_w A_w}{\Delta x}T_w + \frac{k_n A_n}{\Delta y}T_N \quad \Leftrightarrow T_4 = T_P$$

$$\underbrace{\frac{}{}}_{a_P}$$

$$S_P = 0 \qquad S_u = 0$$

- ### Nodes (18,19) :

$$\frac{k_e A_e}{\Delta x}(T_E - T_P) - \frac{k_w A_w}{\Delta x}(T_P - T_W) + \frac{2k_n A_n}{\Delta y}(T_2 - T_P) - \frac{k_s A_s}{\Delta y}(T_P - T_S) = 0$$

$$(\underbrace{\frac{k_e A_e}{\Delta x}}_{a_E} + \underbrace{\frac{k_w A_w}{\Delta x}}_{a_W} + \underbrace{\frac{k_s A_s}{\Delta y}}_{a_S} + \underbrace{\frac{2k_n A_n}{\Delta y}}_{-Sp})T_P = \frac{k_e A_e}{\Delta x}T_E + \frac{k_w A_w}{\Delta x}T_W + \frac{k_s A_s}{\Delta y}T_S + \underbrace{\frac{2k_n A_n}{\Delta y}T_2}_{-Su}$$

$$\underbrace{\frac{}{}}_{a_P}$$

- ### Nodes (06, 07, 10, 11, 14,15)

$$\frac{k_e A_e}{\Delta x}(T_E - T_P) - \frac{k_w A_w}{\Delta x}(T_P - T_W) + \frac{k_n A_n}{\Delta y}(T_N - T_P) - \frac{k_s A_s}{\Delta y}(T_P - T_S) = 0$$

$$(\underbrace{\frac{k_w A_w}{\Delta x}}_{a_W} + \underbrace{\frac{k_e A_e}{\Delta x}}_{a_E} + \underbrace{\frac{k_s A_s}{\Delta y}}_{a_S} + \underbrace{\frac{k_n A_n}{\Delta y}}_{a_N})T_P = \frac{k_w A_w}{\Delta x}T_W + \frac{k_e A_e}{\Delta x}T_E + \frac{k_s A_s}{\Delta y}T_S + \frac{k_n A_n}{\Delta y}T_N$$

$$\underbrace{\frac{}{}}_{a_P}$$

$$S_P = 0 \qquad S_u = 0$$

IV.3.1.3   The third case :

IV.3.1.3.1 Boundary conditions of the "zero flow" type (Neumann), "imposed temperature" (Dirichlet) and "convective flux" (Robin)

- Neumann is an "imposed flow" in our case "null imposed flow" (adiabatic)

It is assumed that the area of the border $x = 0$ and the boundary surface at $x = L$ are maintained at the boundary condition **Neumann** (zero flux) $\left(\dfrac{dT}{dy} = 0 \right)$, the surface of the Border $y = 0$ is maintained at the **Robin** limit condition (convective flow) and the surface of the border $y = H$ is maintained on condition at the **Dirichlet** limit (imposed temperature).

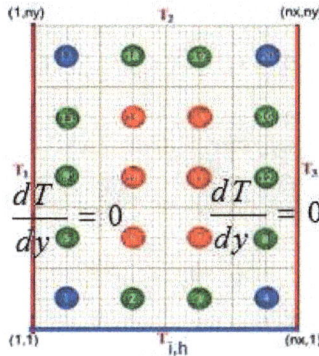

Figure 19. CI control volume Neumann, Dirichlet and Robin

IV.3.1.3.2 Numerical resolution of the heat equation in 2D by the method of finished volumes:

$$\underbrace{\int_v \frac{\partial(\rho T)}{dt} dv}_{\substack{= \\ 0}} + \underbrace{\int_v div\,(\rho T u)dv}_{\substack{= \\ 0}} = \underbrace{\int_v div\,(k\,\overline{gradT}\,)dv}_{\substack{term \\ diffusive}} + \underbrace{\int_v ST\,dv}_{\substack{term \\ source}}$$

$$\int_v div\,(k\,\overline{gradT}\,)dv - \int_v h(T - T_i) = 0 \qquad\qquad (IV\,.6)$$

$$\int_v \frac{\partial}{\partial x}\left(k\,\frac{\partial T}{\partial x} \right)dA_x\,dx + \int_v \frac{\partial}{\partial y}\left(k\,\frac{\partial T}{\partial y} \right)dA_y\,dy - \int_v h(T - T_i) = 0 \qquad (IV\,.7)$$

$$\int_s (k\,\frac{dT}{dx})dA_x + \int_s (k\,\frac{dT}{dy})dA_y - \int_s h \times A_s (T_p - T_i)dy = 0 \qquad (IV\,.8)$$

$$k_e A_e \frac{\partial T}{\partial x}\bigg|_e - k_w A_w \frac{\partial T}{\partial x}\bigg|_w + k_s A_s \frac{\partial T}{\partial y}\bigg|_s - k_n A_n \frac{\partial T}{\partial y}\bigg|_n - \ldots$$

$$\ldots h \times A_s \times dy\,(T_p - T_i) = 0 \tag{IV.10}$$

$$\frac{k_e A_e}{\Delta x}(T_E - T_P) - \frac{k_w A_w}{\Delta x}(T_P - T_W) + \frac{k_n A_n}{\Delta y}(T_N - T_P)\ldots$$

$$\ldots - \frac{k_s A_s}{\Delta y}(T_P - T_S) - h \times A_s \times dy\,(T_p - T_i) = 0 \tag{IV.11}$$

$$a_P T_P = a_W T_W + a_E T_E + a_S T_S + a_N T_N$$

$$a_P = a_W + a_E + a_S + a_N - S_P$$

- **Node (1)**

$$\frac{k_e A_e}{\Delta x}(T_E - T_P) - \underbrace{\frac{2k_w A_w}{\Delta x}}_{0}(T_P - T_1) + \frac{k_n A_n}{\Delta y}(T_N - T_P) - \ldots$$

$$\ldots \frac{2k_s A_s}{\Delta y}(T_P - T_i) - h \times A_s \times dy\,(T_p - T_i) = 0 \quad T_p = T_1$$

$$(\underbrace{\frac{k_e A_e}{\Delta x}}_{a_E} + \underbrace{\frac{k_n A_n}{\Delta y}}_{a_N} + \underbrace{\frac{2k_s A_s}{\Delta y} + h \times A_s \times dy}_{-S_P})T_P = \ldots$$

$$\underbrace{\hspace{6cm}}_{a_P}$$

$$\ldots \frac{k_e A_e}{\Delta x}T_E + \frac{k_n A_n}{\Delta y}T_N + \underbrace{\left(\frac{2k_s A_s}{\Delta y} + h \times A_s \times dy\right)T_i}_{Su}$$

- **Node (4)**

$$\underbrace{\frac{2k_e A_e}{\Delta x}}_{0}(T_3 - T_P) - \frac{2k_w A_w}{\Delta x}(T_P - T_w) + \frac{k_n A_n}{\Delta y}(T_N - T_P) - \frac{2k_s A_s}{\Delta y}(T_P - T_i)\ldots$$

$$\ldots - h \times A_s \times dy\,(T_p - T_i) = 0 \quad T_p = T_3$$

$$(\underbrace{\frac{k_w A_w}{\Delta x}}_{a_W} + \underbrace{\frac{k_n A_n}{\Delta y}}_{a_N} + \underbrace{\frac{2k_s A_s}{\Delta y} + h \times A_s \times dy}_{-S_P})T_P = \frac{k_w A_w}{\Delta x}T_W \ldots$$

$$\underbrace{\hspace{6cm}}_{a_P}$$

$$\ldots + \frac{k_n A_n}{\Delta y}T_n + \left(\frac{2k_s A_s}{\Delta y} + h \times A_s \times dy\right)T_i$$

- **Node (17)**

$$\frac{k_e A_e}{\Delta x}(T_E - T_P) - \frac{2k_w A_w}{\Delta x}(T_P - T_1) + \frac{2k_n A_n}{\Delta y}(T_2 - T_P) - \frac{k_s A_s}{\Delta y}(T_P - T_S) = 0 \quad T_p = T_1$$

$$(\underbrace{\frac{k_e A_e}{\Delta x}}_{a_E} + \underbrace{\frac{k_s A_s}{\Delta y}}_{a_S} + \underbrace{\frac{2k_n A_n}{\Delta y}}_{-S_P})T_P = \frac{k_e A_e}{\Delta x}T_E + \frac{k_s A_s}{\Delta y}T_S + \underbrace{\frac{2k_n A_n}{\Delta y}}_{Su}$$

$$\underbrace{\phantom{(\frac{k_e A_e}{\Delta x} + \frac{k_s A_s}{\Delta y} + \frac{2k_n A_n}{\Delta y})}}_{a_P}$$

- **Node (20)**

$$\underbrace{\frac{2k_e A_e}{\Delta x}(T_3 - T_P)}_{0} - \frac{k_w A_w}{\Delta x}(T_P - T_w) + \frac{2k_n A_n}{\Delta y}(T_2 - T_P) - \frac{k_s A_s}{\Delta y}(T_P - T_s) = 0 \quad T_p = T_3$$

$$(\underbrace{\frac{k_w A_w}{\Delta x}}_{a_W} + \underbrace{\frac{k_s A_s}{\Delta y}}_{a_S} + \underbrace{\frac{2k_n A_n}{\Delta y}}_{-S_P})T_P = \frac{k_w A_w}{\Delta x}T_W + \frac{k_s A_s}{\Delta y}T_S + \underbrace{\frac{2k_n A_n}{\Delta y}}_{Su}$$

$$\underbrace{\phantom{(\frac{k_w A_w}{\Delta x} + \frac{k_s A_s}{\Delta y} + \frac{2k_n A_n}{\Delta y})}}_{a_P}$$

- **Nodes (05, 09,13)**

$$\frac{k_e A_e}{\Delta x}(T_E - T_P) - \frac{2k_w A_w}{\Delta x}(T_P - T_1) + \frac{k_n A_n}{\Delta y}(T_N - T_P) - \frac{k_s A_s}{\Delta y}(T_P - T_s) = 0 \quad T_p = T_1$$

$$(\underbrace{\frac{k_e A_e}{\Delta x}}_{a_E} + \underbrace{\frac{k_s A_s}{\Delta y}}_{a_S} + \underbrace{\frac{k_n A_n}{\Delta y}}_{a_N})T_P = \frac{k_e A_e}{\Delta x}T_E + \frac{k_s A_s}{\Delta y}T_S + \frac{k_n A_n}{\Delta y}$$

$$\underbrace{\phantom{(\frac{k_e A_e}{\Delta x} + \frac{k_s A_s}{\Delta y} + \frac{k_n A_n}{\Delta y})}}_{a_P}$$

$$Su = 0 \qquad Sp = 0$$

- **Nodes (8, 12,16)**

$$\frac{2k_e A_e}{\Delta x}(T_p - T_3) - \frac{k_w A_w}{\Delta x}(T_P - T_w) + \frac{k_n A_n}{\Delta y}(T_N - T_P) - \frac{k_s A_s}{\Delta y}(T_P - T_s) = 0 \quad T_p = T_3$$

$$(\underbrace{\frac{k_w A_w}{\Delta x}}_{a_W} + \underbrace{\frac{k_s A_s}{\Delta y}}_{a_S} + \underbrace{\frac{k_n A_n}{\Delta y}}_{a_N})T_P = \frac{k_w A_w}{\Delta x}T_W + \frac{k_s A_s}{\Delta y}T_S + \frac{k_n A_n}{\Delta y}$$

$$\underbrace{\phantom{(\frac{k_w A_w}{\Delta x} + \frac{k_s A_s}{\Delta y} + \frac{k_n A_n}{\Delta y})}}_{a_P}$$

$$Su = 0 \qquad Sp = 0$$

- ## Nodes (2,3)

$$\frac{k_e A_e}{\Delta x}(T_E - T_P) - \frac{k_w A_w}{\Delta x}(T_P - T_w) + \frac{k_n A_n}{\Delta y}(T_N - T_P) - \frac{2k_s A_s}{\Delta y}(T_P - T_i) - h \times A_s \cdots$$

$$\cdots \times dy\,(T_p - T_i) = 0$$

$$(\underbrace{\underbrace{\frac{k_e A_e}{\Delta x}}_{a_E} + \underbrace{\frac{k_w A_w}{\Delta x}}_{a_W} + \underbrace{\frac{k_n A_n}{\Delta y}}_{a_N} + \underbrace{\frac{2k_s A_s}{\Delta y} + h \times A_s \times dy}_{-S_P})}_{a_P} T_P = \frac{k_e A_e}{\Delta x} T_E + \frac{k_w A_w}{\Delta x} T_W \cdots$$

$$+ \frac{k_n A_n}{\Delta y} T_N + \underbrace{\left(\frac{2k_s A_s}{\Delta y} + h \times A_s \times dy\right) T_i}_{Su}$$

- ## Nodes (18,19)

$$\frac{k_e A_e}{\Delta x}(T_E - T_P) - \frac{k_w A_w}{\Delta x}(T_P - T_w) + \frac{2k_n A_n}{\Delta y}(T_2 - T_P) - \frac{2k_s A_s}{\Delta y}(T_P - T_S) = 0$$

$$(\underbrace{\underbrace{\frac{k_e A_e}{\Delta x}}_{a_E} + \underbrace{\frac{k_w A_w}{\Delta x}}_{a_W} + \underbrace{\frac{k_s A}{\Delta y}}_{a_S} + \underbrace{\frac{2k_n A_n}{\Delta y}}_{-S_P})}_{a_P} T_P = \frac{k_e A_e}{\Delta x} T_E + \frac{k_w A_w}{\Delta x} T_W + \frac{k_s A_s}{\Delta y} T_S + \underbrace{\frac{2k_n A_n}{\Delta y} T_2}_{Su}$$

- ## Nodes (6,7 ,10 ,11 ,14 ,15)

$$\frac{k_e A_e}{\Delta x}(T_E - T_P) - \frac{k_w A_w}{\Delta x}(T_P - T_W) + \frac{k_n A_n}{\Delta y}(T_N - T_P) - \frac{k_s A_s}{\Delta y}(T_P - T_S) = 0$$

$$(\underbrace{\underbrace{\frac{k_w A_w}{\Delta x}}_{a_W} + \underbrace{\frac{k_e A_e}{\Delta x}}_{a_E} + \underbrace{\frac{k_s A_s}{\Delta y}}_{a_S} + \underbrace{\frac{k_n A_n}{\Delta y}}_{a_N})}_{a_P} T_P = \frac{k_w A_w}{\Delta x} T_W + \frac{k_e A_e}{\Delta x} T_E + \frac{k_s A_s}{\Delta y} T_S + \frac{k_n A_n}{\Delta y} T_N$$

$$Su = 0 \qquad Sp = 0$$

IV.4 MATLAB Code Flowchart :

To solve our problem, we used the **MATLAB** version calculation **code** R1015a.

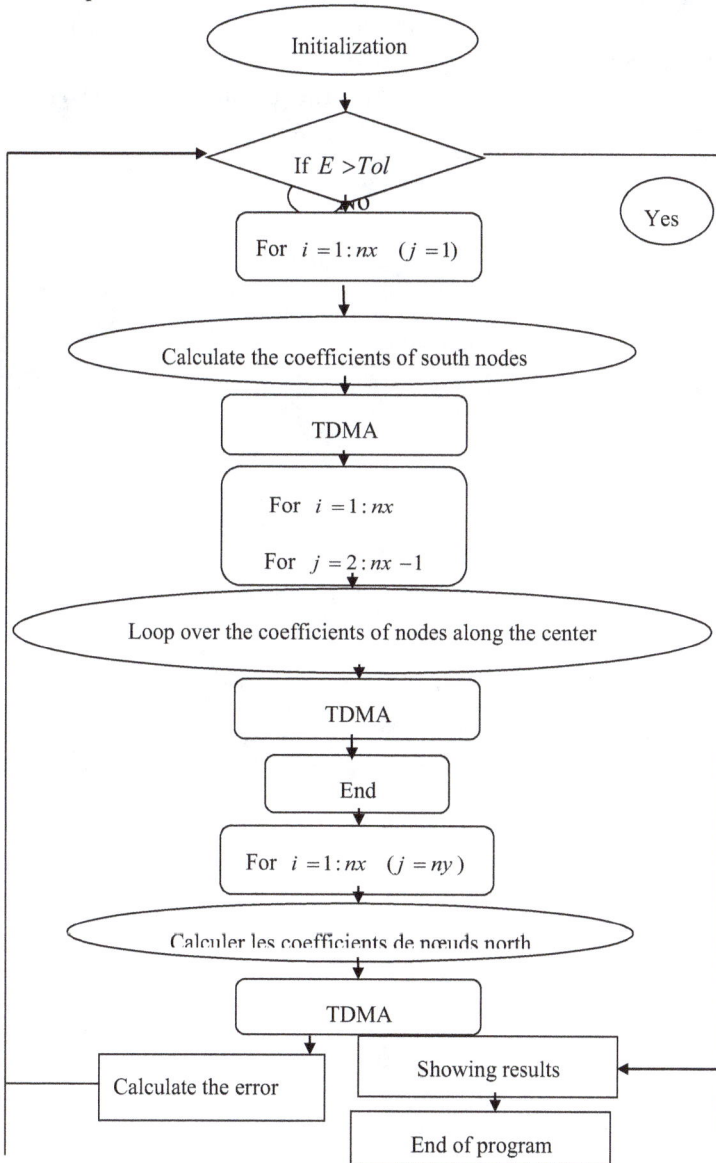

Figure 20. program flowchart

IV.5 Conclusion :

In this chapter, we have presented a two-dimensional numerical solution of the governing equation by the finite volume method which is presented in the previous chapter with three cases of boundary
condition **Neumann**, **Dirichlet** and **Robin**, The resolution is carried out by **MATLAB** code.

Chapter V:

Results and interpretations

V.   Chapter V. Results and interpretations

V.1   Introduction :

This chapter includes the comparison of the numerical results of **MATLAB** programs and **COMSOL**, by employing four methods of the resolution of the matrix: TDMA, Gausse- Seidel without and with relaxation and Jacobi on **MATLAB**.

V.2   Mesh test:

We made the mesh test to know the minimum threshold of the mesh for the appearance of the temperature distribution; we came to the following results

nodes	4x5	8x10	16x20	32x40	64x80
1	39,13	42,41	44,41	45,32	45,77
2	62,30	63,10	63,56	63,82	63,95
3	74,44	74,16	74,13	74,17	74,19
4	79,72	79,26	79,50	79,72	79,85

Tableau 2. Mesh test

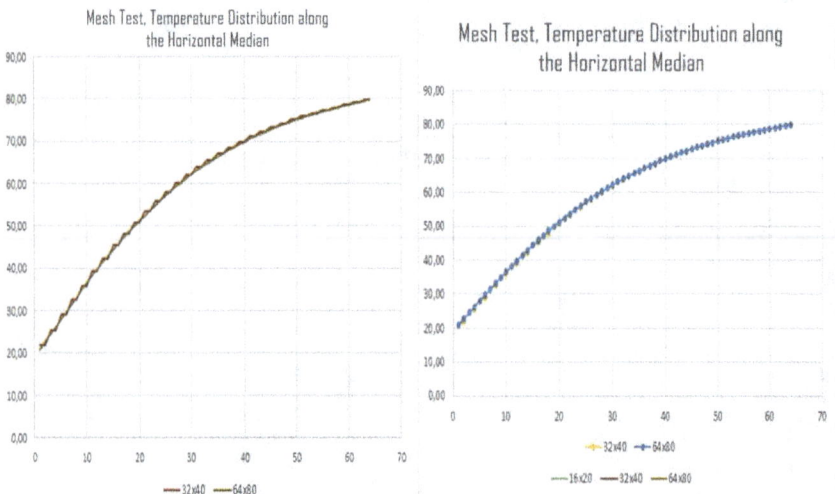

Mesh Test, Temperature Distribution along the Horizontal Median

Mesh Test, Temperature Distribution along the Horizontal Median

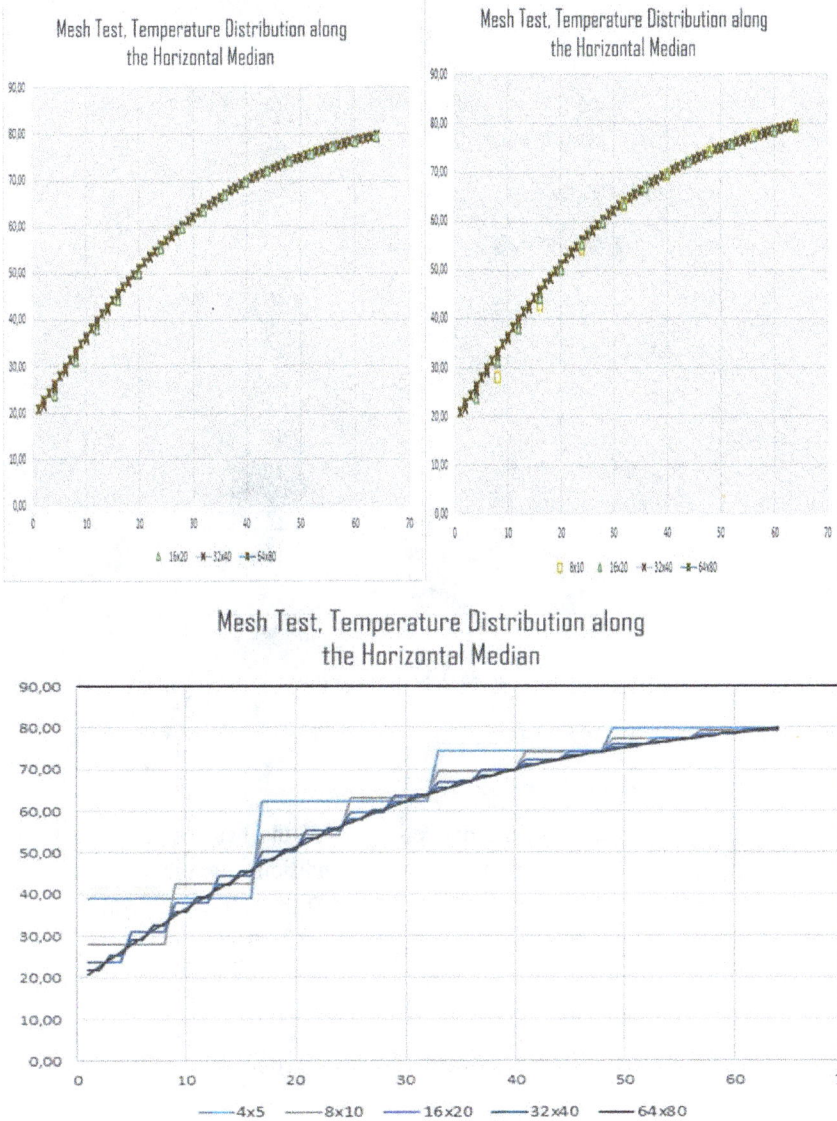

Figure 21. Mesh test curves

V.2.1   Discussions:

Figure 21 represents the mesh test curves, according to the test, it was concluded that the mesh (nx = 32 ny = 40) is the minimum threshold for the variation of the distribution of the temperature along the median.

V.1   Data initialization:

For the validation of our results, we took the following data for each case:

the data	Lx	Ly	dz	T_1	T_2	T_3	T_4	K	h	error	iteration
1st case	0,4	0.5	0.001	40	60	80	100	100		10^{-5}	5000
2nd case	0,4	0.5	0.001	40	60	80	100	100		10^{-5}	5000
3rd case	0,4	0.5	0.001	40	60	80	100	100	5	10^{-5}	5000

Tableau 3. Data initialization

For the third case, we took $T_4=T_i$

V.2   Validation of our results:

The main aim of our study is the discretization of the heat equation in 2D in a rectangular plate, with different boundary conditions by the finite volume method. According to our calculations we arrived at the following results:

V.2.1   By TDMA

V.2.1.1   The first case

In the first case we have discretized with a boundary condition of the **Dirichlet** type:

MATLAB   **COMSOL**

- The results obtained with (nx = 4 ny = 5)

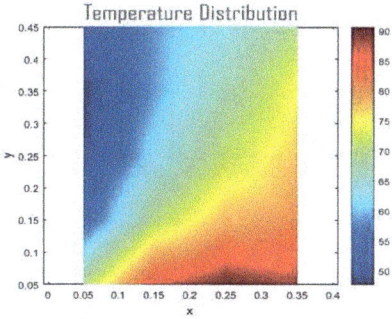
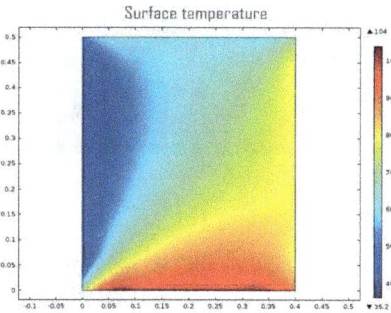

(a)   (b)

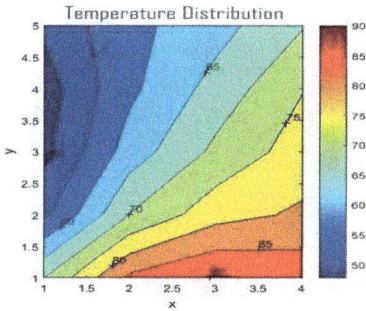
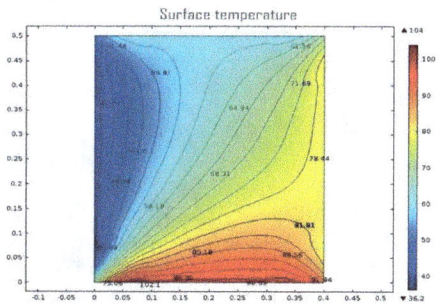

(c)   (d)

*Figure 22. The distribution and isotherms of temperature with a **dirichlet** boundary condition*

(nx=4 ny=5) method TDMA

- The results obtained with (nx=40 ny=50)

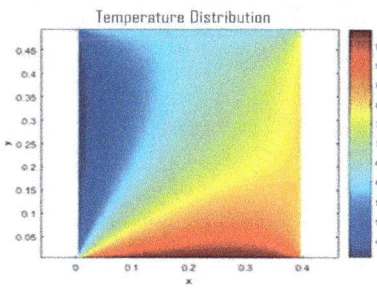
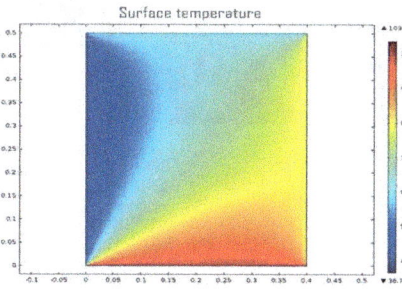

(a)   (b)

(c)   (d)

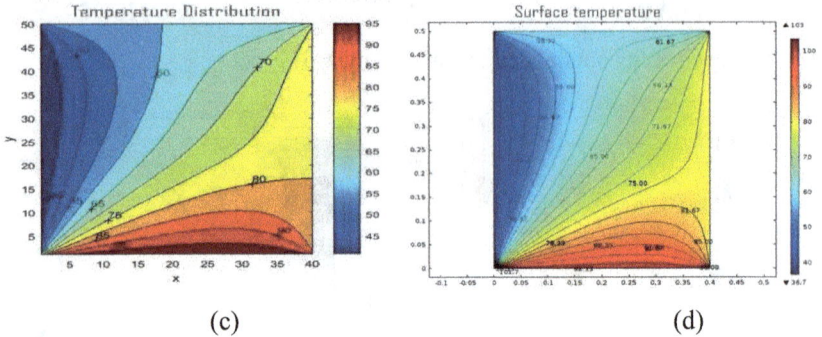

*Figure 23.The distribution and isotherms of temperature with a **dirichlet** boundary condition*

(nx=40 ny=50) method TDMA

- The results obtained with (nx=80 ny=100)

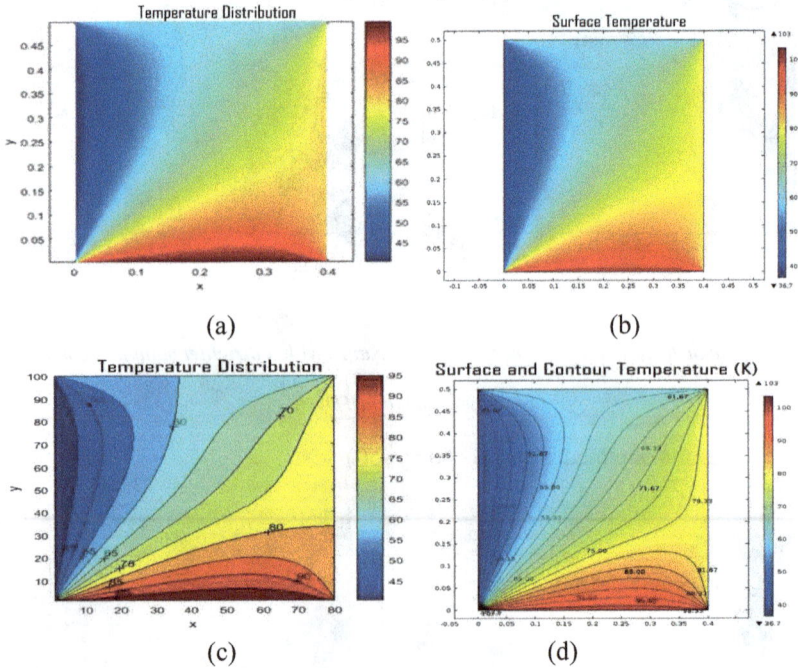

(a)   (b)

(c)   (d)

*Figure 24. The distribution and isotherms of temperature with a **dirichlet** boundary condition*

(nx=80 ny=100) method TDMA

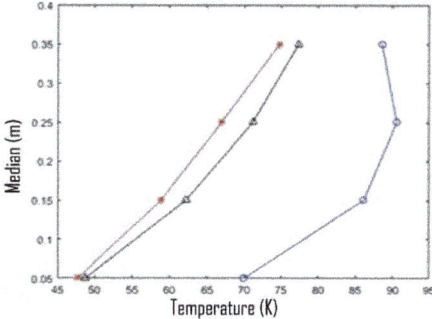
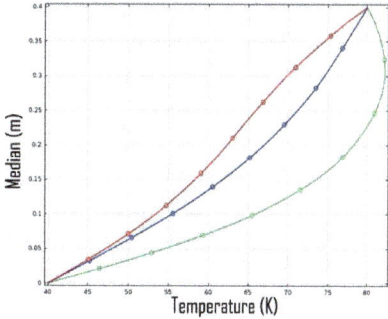

a1   (nx=4 ny=5)   a2

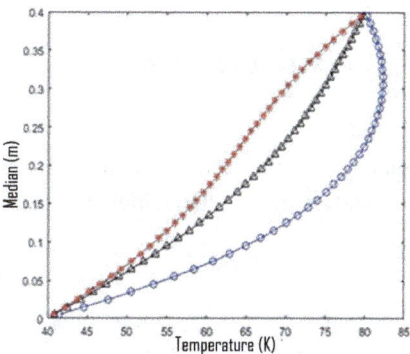
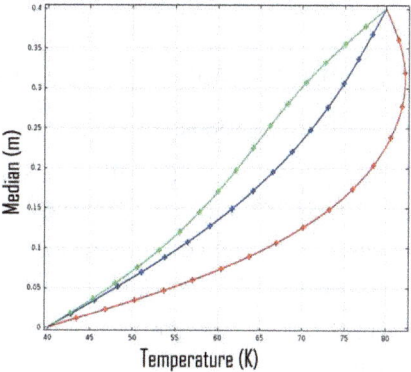

b1   (nx=40 ny=50)   b2

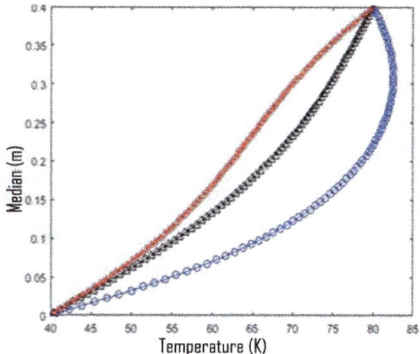
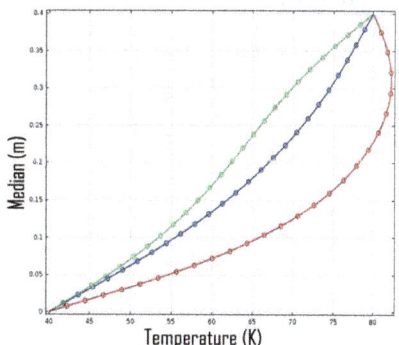

c1   (nx=80 ny=100)   c2

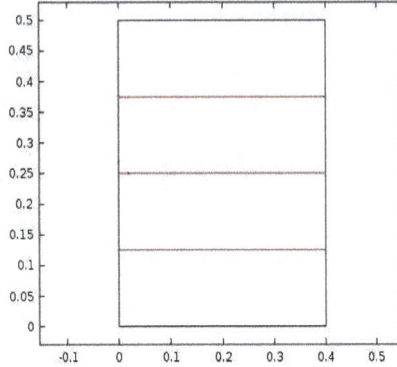

Figure 25. (a1, a2, b1, b2, c1 and c2) Represents the distribution of temperature, along the three

*rows: the median, above and below the median by **MATLAB** and **COMSOL***

V.2.1.1.1 Discussions:

Figures 22, 23 and 24 show the comparison of the numerical results between the code of **MATLAB** and **COMSOL** computation in the **Dirichlet** boundary condition, one notices a agreement between the results obtained by the two codes, except for a coarse mesh (4x5) we notice inconsistent results.

Figure 25 represents the temperature distribution, along the three lines: the median, the above and below the median **MATLAB** and **COMSOL**, one also notices a similarity between the results obtained, except for a coarse mesh (4x5) one notices results which are not similar.

V.2.1.2 The second case

In the second case one discretizes with the mixed boundary conditions of the **Dirichlet** type and **Neumann**:

MATLAB   **COMSOL**

- The results obtained with (nx=4 ny=5)

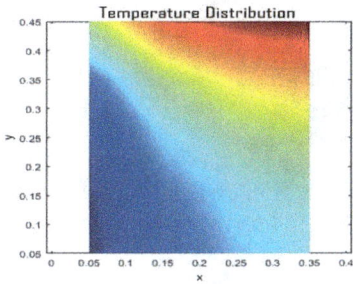
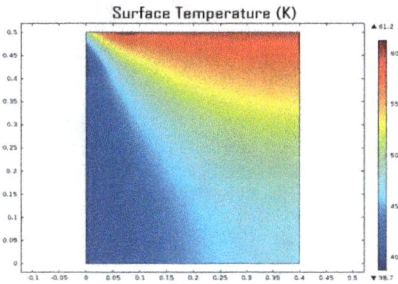

(a)   (b)

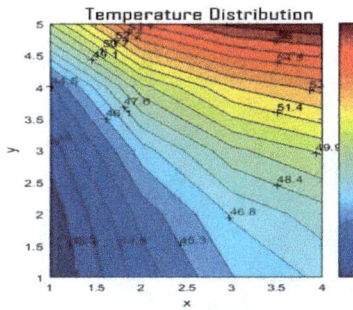
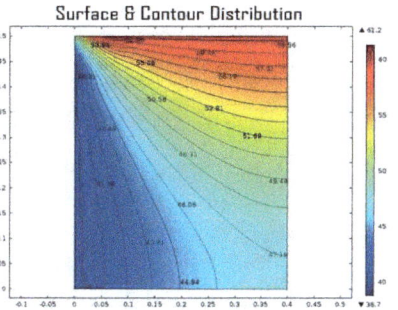

(c)   (d)

*Figure 26. The distribution and isotherms of temperature with the **dirichlet** and **Neumann** boundary condition (nx=4 ny=5) method TDMA*

- The results obtained with (nx=40 ny=50)

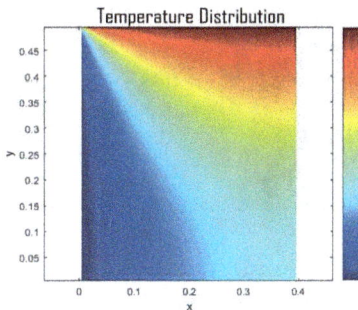
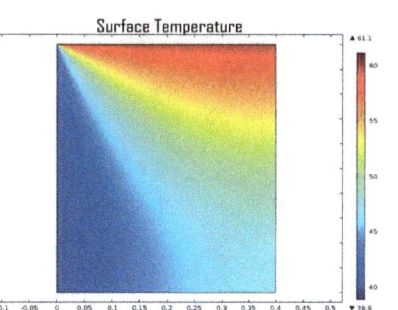

(a)   (b)

(c)   (d)

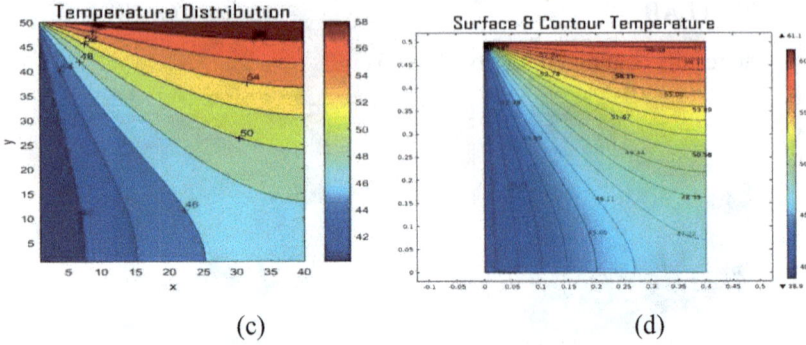

*Figure 27. The distribution and isotherms of temperature with the **dirichlet** and **Neumann** boundary condition (nx=40 ny=50) method TDMA*

• The results obtained with (nx=80 ny=100)

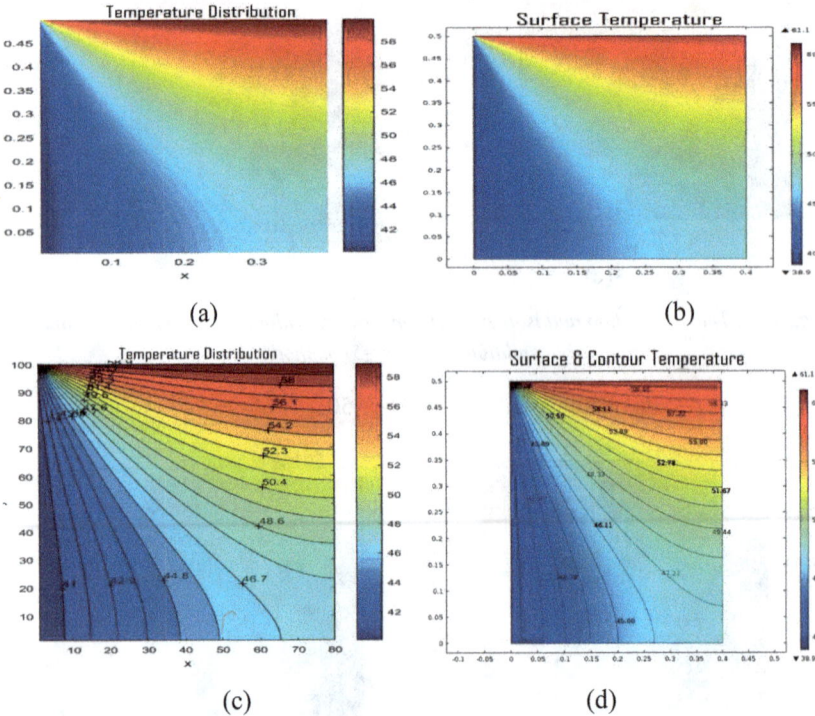

(a)   (b)

(c)   (d)

*Figure 28. The distribution and isotherms of temperature with the **dirichlet** and **Neumann** boundary condition (nx=80 ny=100) method TDMA*

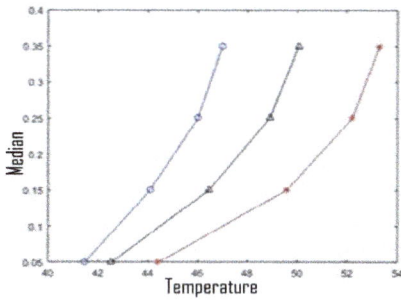 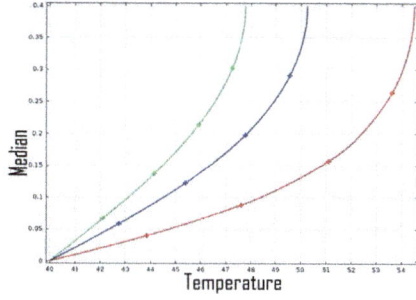

a1   (nx=4 ny=5)   a2

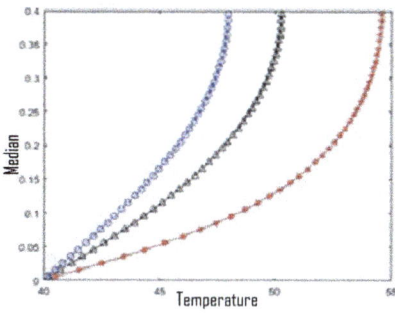 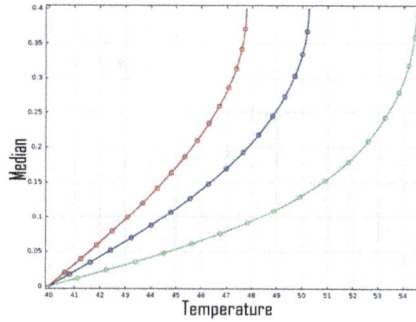

b1   (nx=40 ny=50)   b2

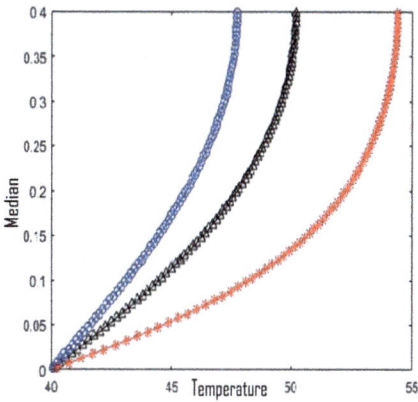 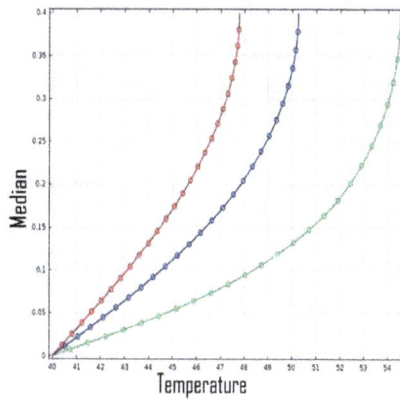

c1   (nx=80 ny=100)   c2

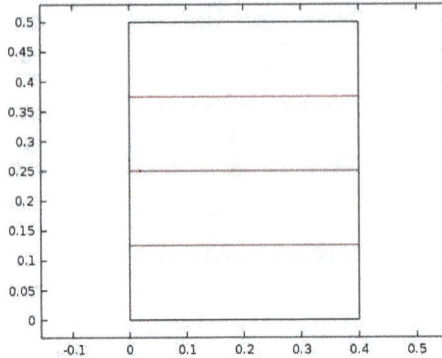

*Figure 29. (a1, a2, b1, b2, c1 and c2) Represents the distribution of temperature, along the three rows: the median, above and below the median by **MATLAB** and **COMSOL***

V.2.1.2.1 Discussions

Figures 26,27 and 28 represent the comparison of the numerical results between the code of calculates MATLAB and COMSOL under the boundary conditions Dirichlet and Neumann, one also notices a similarity between the results obtained by the two codes, except for a coarse mesh (4x5) one notices non-concordant results.

A certain instability compared to COMSOL, is very noticed in the cases of isothermal, this instability disappears for a finer mesh.

Figure 29 represents the temperature distribution, along the three lines : the median, the above and below the median MATLAB and COMSOL, one also notices a agreement between the results obtained, except for a coarse mesh (4x5) the non similar .

V.2.1.3 Third case

In the Third case one discretizes with the mixed boundary conditions of the **Dirichlet** type, **Neumann** and **Robin**:

MATLAB   COMSOL

- The results obtained with (nx=4 ny=5)

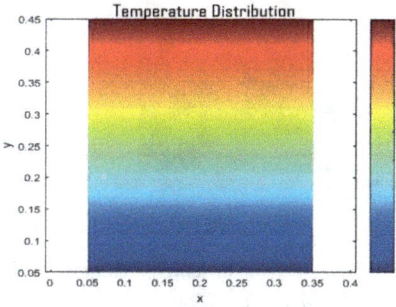

(a)

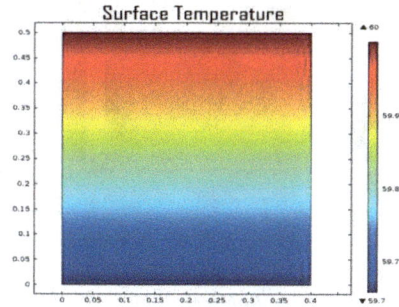

(b)

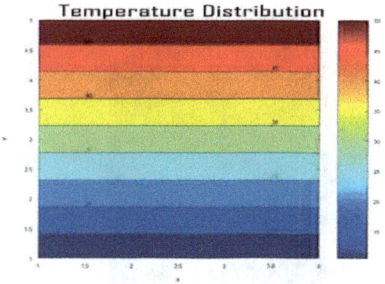

(c)

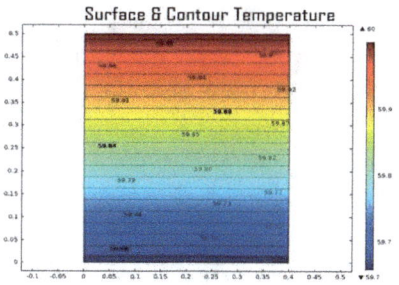

(d)

*Figure 30. The distribution and isotherms of temperature with the **dirichlet**, **Neumann** and **Robin** boundary condition (nx=4 ny=5) method TDMA*

- The results obtained with (nx=40 ny=50)

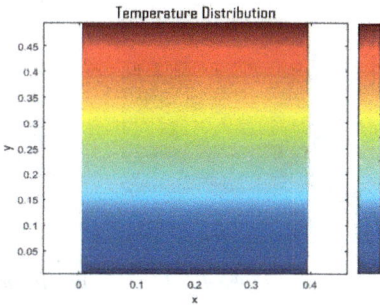

(a)

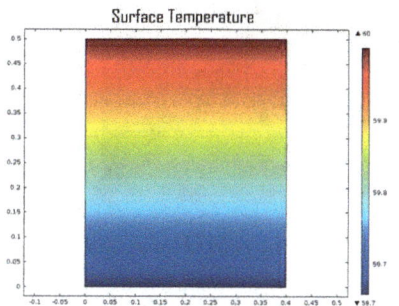

(b)

(c)   (d)

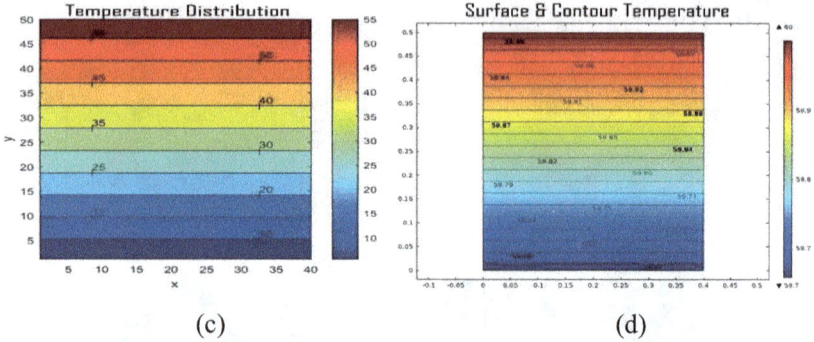

*Figure 31. The distribution and isotherms of temperature with the **dirichlet, Neumann** and*
***Robin** boundary condition (nx=40 ny=50) method TDMA*

- The results obtained with (nx=80 ny=100)

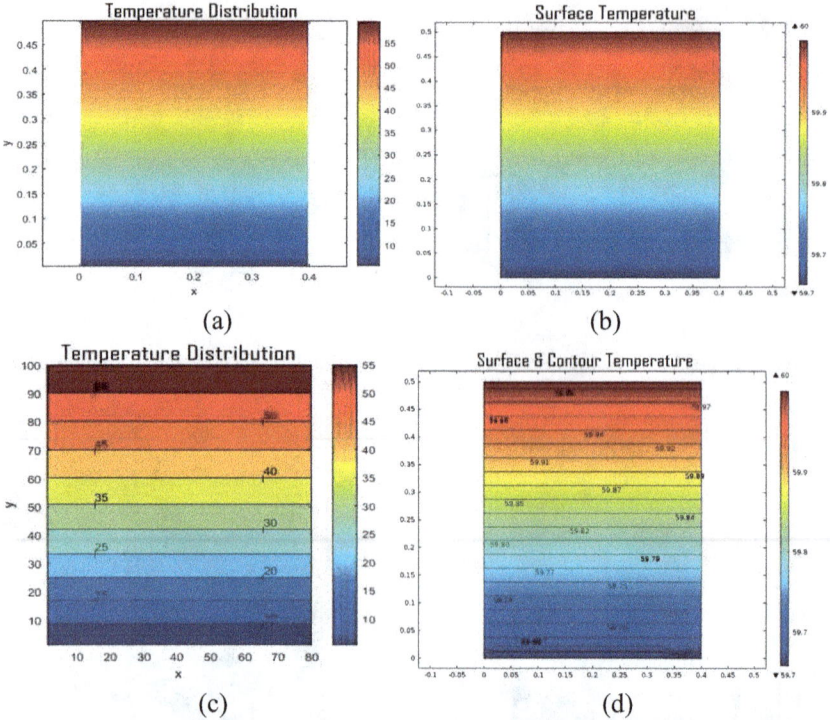

(a)   (b)

(c)   (d)

*Figure 32. The distribution and isotherms of temperature with the **dirichlet, Neumann** and*
Robin boundary condition (nx=80 ny=100) method TDMA

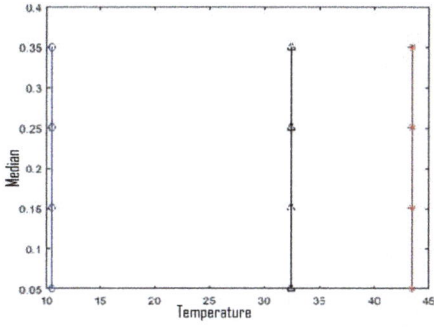
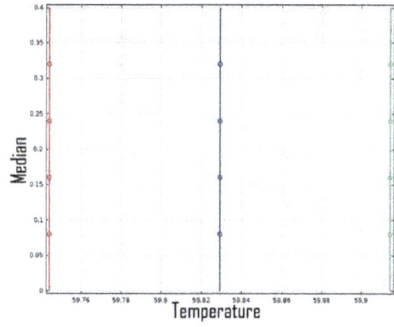

a1   (nx=4 ny=5)   a2

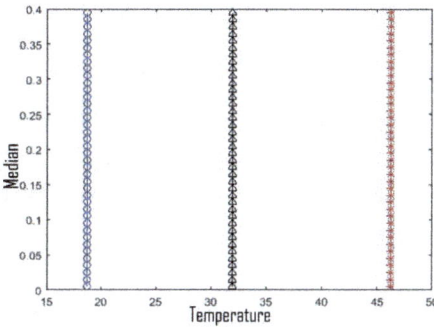
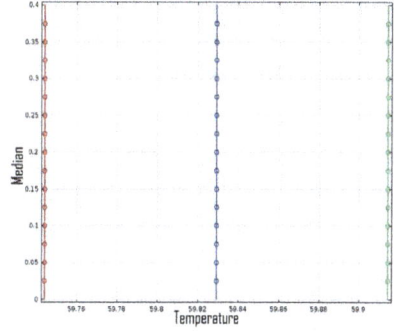

b1   (nx=40 ny=50)   b2

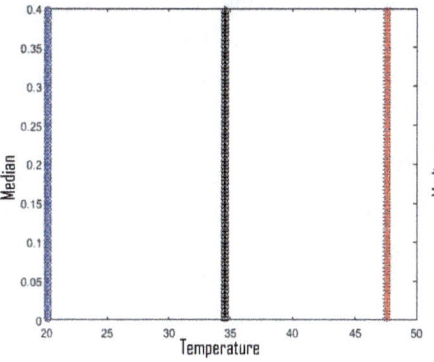
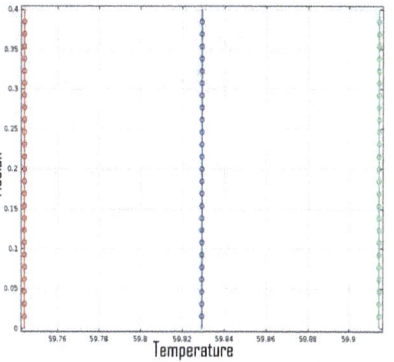

c1   (nx=80 ny=100)   c2

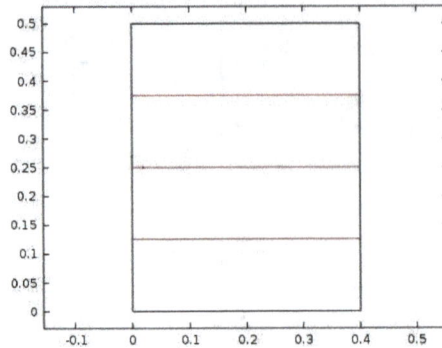

*Figure 33. (a1, a2, b1, b2, c1 and c2) Represents the distribution of temperature, along the three rows: the median, above and below the median by **MATLAB** and **COMSOL***

V.2.1.3.1 Discussions

Figures 30, 31 and 32 represent the comparison of the numerical results between the code of computes **MATLAB** and **COMSOL** under the **Dirichlet Newman** boundary conditions and **Robin**, one also notices similar results even by a coarse mesh, and a similarity between the results obtained between the two codes.

A certain instability compared to **COMSOL**, is very noticed in the cases of isothermal, this instability disappears for a finer mesh.

Figure 33 represents the temperature distribution, along the three lines: the median, the above and below the median **MATLAB** and **COMSOL**, we also note closer results for a finer mesh, and for a coarse mesh (4x5) the results are not similar.

V.2.2   The methods of resolution of the matrix

During the discretions of the heat equation by the FVM method one obtained a tridiagonal hollow matrix. this matrix is usually solved by the TDMA method, but we wanted to solve it with other classical methods used for plain matrices. To see the speed of the execution of the code, by these methods, we carried out simulations for a mesh of (80x100):

- Gauss- Seidel Without relaxation
- Gauss- Seidel with relaxation
- Jacobi

And to know more information on these methods you can see the appendices, pages (pp 74-78).

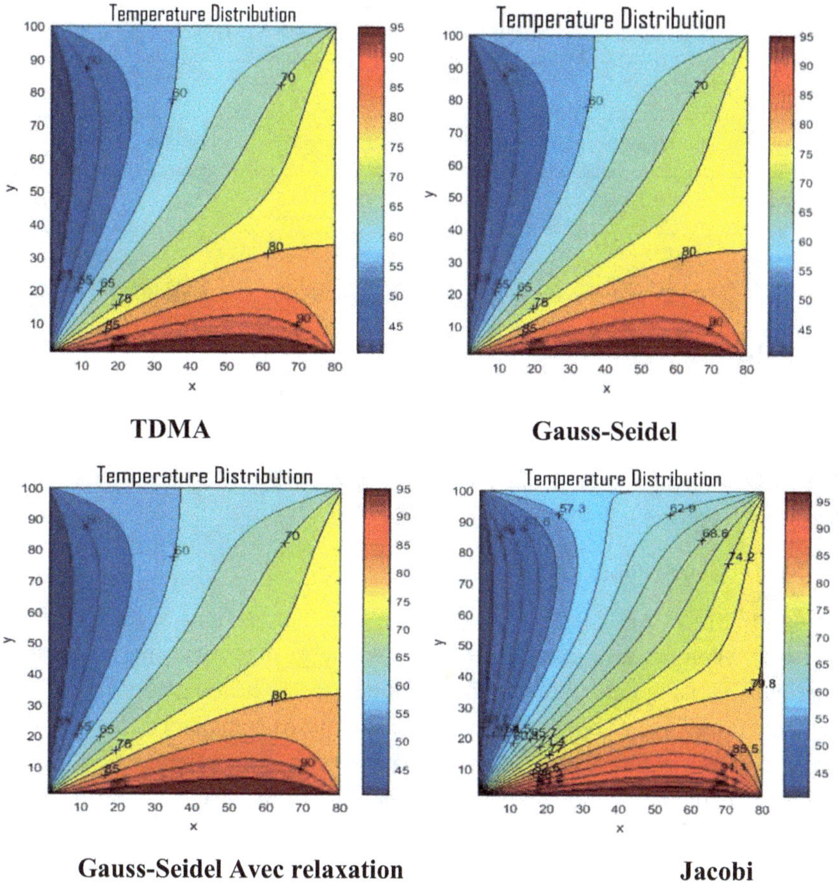

TDMA

Gauss-Seidel

Gauss-Seidel Avec relaxation

Jacobi

*Figure 34. The temperature isotherms with a **Dirichlet** boundary condition by the four methods*

V.2.2.1   Discussions

Figure 34 represents the isotherms of the temperature with **Dirichlet** boundary condition by the four methods, we notice similar results, we see the difference between the results executed at the level of the number of iterations (execution time).

V.3   Comparison table of iterations between the four methods

Tolerance	Mesh	TDMA	GSSR	GSR	Jacobi
1×10^{-4}	4×5	826	826	826	826
1×10^{-4}	40 ×50	1873	1872	1875	1869
1×10^{-4}	80×100	1954	1957	1959	1968

Tableau 4. Comparison of iterations between the four methods

V.3.1.1   Discussions

Table 4 represents the comparison of the iterations between the four methods; we notice the number of iterations closest for all methods.

The TDMA method remains the most reliable method, and especially for a tight mesh, it converges quickly

V.4   Conclusion:

Using the MATLAB language, we wrote a code by the finite volume method, which is a conservative method; it was possible to determine the temperature distributions, isotherms, the numerical solution of the heat equation, as well as the validation with a known commercial code known commercial, based on the finite element method, we compared our results for different mesh, using different methods of solving the tridiagonal matrix. We have noticed that the results obtained were very satisfactory.

Generale conclusion

Generale conclusion

The work carried out as part of this thesis is intended to better understand a study digital heat transfer by conduction in a rectangular plate, based on a numerical study based on the discretization of the 2D heat equation by the finite volume method, The boundary conditions are of the **Dirichlet** type (imposed temperature, **Neumann** (zero heat flow) and **Robin** (convective flow).

The treated subject allowed us to know and to implement several digital methods:

- The numerical resolution and discretization of the heat equation by the most conservative method, finite volume method (FVM).
- Method of solving the linear equation system, TDMA, Gauss-Seidel with and without relaxation, as well as Jacobi.

The results of numerical simulations are obtained with a calculation code that we have developed and tested by **MATLAB**, the results obtained are compared with **COMSOL**.

From the results obtained we conclude:

- The mesh (nx = 32 ny = 40) is the minimum threshold for the variation of the distribution of the temperature along the median.
- Q concordance between the results obtained by the two codes except for a coarse mesh (4x5) we notice inconsistent results, for the three cases.
- Similar results between the four methods, we can see the difference between the results executed at the iteration count (execution time) level.
- The variation of the iterations in the mesh (nx = 4 ny = 5), but for a finer mesh the closest iterations for all methods except Jacobi.
- The TDMA method remains the most reliable method, because it is the fastest to Converge.

In future work, we suggest expanding this work by:

Solving the convection-diffusion equation, and that of Navier-Stokes by the method finite volumes in a shifted mesh, we have used the three variables (u, v, p), and validated the results obtained with the benchmarks of Vahl Davis , and of Ghia of a differentially heated square cavity, a cavity with a moving wall (Lid driven cavity).

Bibliographic references

Bibliographic references

[1] M. T. Manzari and M. T. Manzari, "ON Numerical solution of hyperbolic heat conduction," *numerical method in biomedical engineering* .,vol. 866, no. October 1998, pp. 853–866, 1999.

[2] A. Grine, J. Y. Desmons, and S. Harmand, "Models for transient conduction in a flat plate subjected to a variable heat flux," *Appl. Therm. Eng.*, vol. 27, no. 2–3, pp. 492–500, 2007.

[3] A. Gersborg-Hansen, M. P. Bendsøe, and O. Sigmund, "Topology optimization of heat conduction problems using the finite volume method," *Struct. Multidiscip. Optim.*, vol. 31, no. 4, pp. 251–259, 2006.

[4] A. V. Itagi, "Finite volume method for the fourier heat conduction in layered media with a moving volume heat source," *Japanese J. Appl. Physics, Part 1 Regul. Pap. Short Notes Rev. Pap.*, vol. 46, no. 4 A, pp. 1482–1489, 2007.

[5] J. T. S. A.Diószegi, É. Diószegi, J.Tóth, "Modelling and simulation of heat conduction in 1-D polar spherical coordinates using control volume-based finite difference method," *Int. J. Numer. Methods Heat Fluid Flow*, vol. 26, no. 1, pp. 2–17, 2015.

[6] G. Sachdeva, K. S. Kasana, and R. Vasudevan, "Heat transfer enhancement by using a rectangular wing vortex generator on the triangular shaped fins of a plate-fin heat exchanger," *Heat Transf. - Asian Res.*, vol. 39, no. 3, pp. 151–165, 2010.

[7] S. Mazumder, "Comparative Assessment of the Finite Difference, Finite Element, and Finite Volume Methods for a Benchmark One-Dimensional Steady-State Heat Conduction Problem," *J. Heat Transfer*, vol. 139, no. 7, p. 71301, 2017.

[8] Q. Xue, X. F. Xiao, and N. Z. Fan, "Heat Conduction Equation Finite Volume Method to Achieve on MATLAB," *Mech. Eng. Intell. Syst. Pts 1 2*, vol. 195–196, pp. 712–717, 2012.

[9] S. Murakami and Y. Asako, "A finite volume method on distorted quadrilateral meshes for discretization of the energy equation's conduction term," *Heat Transf. - Asian Res.*, vol. 42, no. 2, pp. 163–184, 2013.

[10] T. Lee, M. Leok, and N. H. McClamroch, "Geometric numerical integration for complex dynamics of tethered spacecraft," *Proc. 2011 Am. Control Conf.*, vol. m, no.

March, pp. 1885–1891, 2011.

[11]   W. Li, B. Yu, X. Wang, P. Wang, and S. Sun, "A finite volume method for cylindrical heat conduction problems based on local analytical solution," *Int. J. Heat Mass Transf.*, vol. 55, no. 21–22, pp. 5570–5582, 2012.

[12]   P. Duda, "Finite element method formulation in polar coordinates for transient heat conduction problems," *J. Therm. Sci.*, vol. 25, no. 2, pp. 188–194, 2016.

[13]   O. N. T. H. E. Effect *et al.*, "A Riemann-Hilbert problem for a heat conduction in an infinite plate with a rectangular hole," *Int. J. Heat Mass Transf.*, vol. 16, no. 10, pp. 1941–1943, 1973.

[14]   H. T. Kim, B. W. Rhee, and J. H. Park, "Application of the finite volume method to the radial conduction model of the CATHENA code," *Ann. Nucl. Energy*, vol. 33, no. 10, pp. 924–931, 2006.

[15]   R. Chaabane, F. Askri, and S. Ben Nasrallah, "Analysis of two-dimensional transient conduction-radiation problems in an anisotropically scattering participating enclosure using the lattice Boltzmann method and the control volume finite element method," *Comput. Phys. Commun.*, vol. 182, no. 7, pp. 1402–1413, 2011.

[16]   C. Luo, B. Moghtaderi, and A. Page, "Modelling of wall heat transfer using modified conduction transfer function, finite volume and complex Fourier analysis methods," *Energy Build.*, vol. 42, no. 5, pp. 605–617, 2010.

[17]   S. Han, "Finite volume solution of two-step hyperbolic conduction in casting sand," *Int. J. Heat Mass Transf.*, vol. 93, pp. 1116–1123, 2016.

[18]   S. Singh, P. K. Jain, and Rizwan-uddin, "Analytical solution to transient heat conduction in polar coordinates with multiple layers in radial direction," *Int. J. Therm. Sci.*, vol. 47, no. 3, pp. 261–273, 2008.

[19]   J. J. J.Stefaniak, "AN APPROXIMATE SOLUTION OF HEAT CONDUCTION EQUATION WITH MIXED," *Mech. Res. Commun.*, vol. 25, no. 6, pp. 631–636, 1998.

[20]   A. Evgrafov, M. M. Gregersen, and M. P. Sorensen, "Convergence of cell based finite volume discretizations for problems of control in the conduction coefficients," *ESAIM Math. Model. Numer. Anal.*, vol. 45, no. 6, pp. 1059–1080, 2011.

[21] S. Han, "Finite volume solution of a 1-D hyperbolic conduction equation," *Numer. Heat Transf. Part A Appl.*, vol. 67, no. 5, pp. 497–512, 2015.

[22] Z.Abdallah, "DETERMINATION DU CHAMP DE TEMPERATURES DANS UNE CAVITE PLEINE EN UTILISANT LA METHODE MIXTE VOLUMES FINIS – ELEMENTS FINIS Soutenu," UNIVERSITE DE OUARGLA, 2004.

[23] K. D. L.Hafnaoui, "Optimisation du bilan thermique d'un circuit de refroidissement du solvant lourd en présence du phénomène d'encrassement," UNIVERSITÉ ECHAHID HAMMA LAKHDAR EL OUED, 2015.

[24] J.Brau, *Transferts de chaleur et de masse . INSA .LYON.* 2006.

[25] H.abdelkrim, *Transferts Thermique.* Dar-El-Djazairia.Alger. 2001.

[26] M. Djamel, "Étude de l'influence des paramètres climatiques sur la température du sol (application au site de Biskra)," Université de Biskra Faculté, 2013.

[27] W. KORICHI, "Simulation numérique d'une plaque bidimensionnelle avec source de chaleur," Université Mohamed khider – BISKRA –, 2014.

[28] K. Fayçal, "Conversion thermodynamique de l'énergie solaire: Etude et modélisation d'un capteur solaire," UNIVERSITE DE BATNA, 2014.

[29] M.Gacem, "Comparaison Entre l'Isolation Thermique Extérieure et Intérieure d'une pièce D'un Habitat Situé Dans Le Site De Ghardaïa," L'UNIVERSITE ABOU-BEKR BELKAID-TLEMCEN, 2011.

[30] E.Goncalvès, *Méthodes et Analyse Numériques.* Engineering school. Institut Polytechnique de Grenoble. 2007.

[31] N.Ouassila, "Etude et modélisation des paliers planaires," UNIVERSITE MENTOURI DE CONSTANTINE, 2009.

[32] H.Karima, "ETUDE DU TRANSFERT DE CHALEUR A TRAVERS UNE AILETTE VERTICALE," de constantine.

[33] M.Younes, "Etude Numérique Comparative Entre Deux Types de Chicanes et Ailettes (Trapézoïdale et Triangulaire) Utilisées Pour Améliorer les Performances des Capteurs Solaires Plans à Air.''," Abou Bekr Belkaïd Tlemcen, 2012.

Appendices

Appendices

Appendices A: MATLAB code

MATLAB (matrix laboratory) is a fourth generation programming language emulated by a development environment of the same name, it is used for of numerical computation. Developed by The MathWorks company, **MATLAB** allows you to manipulate matrices, display curves and data, implement algorithms, to create user interfaces.

• Example of calculation of the coefficients of the south nodes

```
% Noeud SE   (4)       %----noeud---4---SE-------------
    aw(nx,1)=kw*Aw/dx;
    ae(nx,1)=0.;
    as(nx,1)=0.;
    an(nx,1)=kn*An/dy;
    Sp(nx,1)=-(2*ke*Ae/dx + 2*ks*As/dy);
    Su(nx,1)=2*ke*Ae/dx*T3+ 2*ks*As/dy*T4;
    ap(nx,1)= an(nx,1)+ aw(nx,1)+ as(nx,1)+ ae(nx,1)- Sp(nx,1);
    a(nx)=-aw(nx,1);
    b(nx)=ap(nx,1);
    c(nx)=-ae(nx,1);
    d(nx)=Su(nx,1) + an(nx,1)*Told(nx,2);
    erreur=erreur+ abs(an(nx,1)*Told(nx,2)+aw(nx,1)*T(nx-1,1)+Su(nx,1)-ap(nx,1)*T(nx,1));
    Frp = Frp + abs(ap(nx,1)*T(nx,1));
%----fin cellule Sud--1----2----3----4------
    %-------TDMA--------------------
    T(:,1)= MyTDMA(a,b,c,d);
```

• Example of calculation of the coefficients of the nodes of center

```
%----noeuds intr---6--7--10--11--14--15---
for i=2:nx-1
    aw(i,j)=kw*Aw/dx;
    ae(i,j)=ke*Ae/dx;
    as(i,j)=ks*As/dy;
    an(i,j)=kn*An/dy;
    Sp(i,j)=0.;
    Su(i,j)=0.;
    ap(i,j)=as(i,j)+ an(i,j)+ ae(i,j)+ aw(i,j)- Sp(i,j);
    a(i)=-aw(i,j);
    b(i)=ap(i,j);
    c(i)=-ae(i,j);
    d(i)=Su(i,j) + an(i,j)*Told(i,j+1)+as(i,j)*T(i,j-1);
    erreur=erreur+abs(ae(i,j)*Told(i+1,j)+an(i,j)*Told(i,j+1)+as(i,j)*T(i,j-1)+aw(i,j)*T(i-1,j)+Su(i,j)-ap(i,j)*T(i,j));
    Frp = Frp + abs(ap(i,j)*T(i,j));
end
```

Appendix B: TDMA method

The Tridiagonal Matrix Algorithm (TDMA), also known as the Thomas Algorithm, is a simplified form of Gaussian elimination that can be used to solve systems of tridiagonal equations. A tridiagonal system can be written as:

$$a_i x_{i-1} + b_i x_i + c_i x_{i+1} = d_i$$

Ou $a_1 = 0$ et $c_n = 0$

$$\begin{bmatrix} b_1 & c_1 & & & & 0 \\ a_2 & b_2 & c_2 & & & \\ & a_3 & b_3 & \ddots & & \\ & & \ddots & \ddots & c_{n-1} \\ 0 & & & a_n & b_n \end{bmatrix} \begin{bmatrix} x_1 \\ x_2 \\ x_3 \\ \vdots \\ x_n \end{bmatrix} = \begin{bmatrix} d_1 \\ d_2 \\ d_3 \\ \vdots \\ d_n \end{bmatrix}$$

Exemple:

$$\begin{bmatrix} a_p & -a_w & 0 & 0 \\ -a_e & a_p & -a_w & 0 \\ 0 & -a_e & a_p & -a_w \\ 0 & 0 & -a_e & a_p \end{bmatrix} \begin{bmatrix} T_1 \\ T_2 \\ T_3 \\ T_4 \end{bmatrix} = \begin{bmatrix} Su_1 + a_s T_s + a_n T_n \\ Su_2 + a_s T_s + a_n T_n \\ Su_3 + a_s T_s + a_n T_n \\ Su_4 + a_s T_s + a_n T_n \end{bmatrix}$$

Algorithm:

```
function Phi = MyTDMA(a,b,c,d)
nx=length(a);
for k=2:nx
    m=a(k)/b(k-1);
    b(k)=b(k)-m*c(k-1);
    d(k)=d(k)-m*d(k-1);
end
Phi(nx)=d(nx)/b(nx);
%....phase de substitution en arriere....%
for k=nx-1:-1:1
    Phi(k)=(d(k)-c(k)*Phi(k+1))/b(k);
end
```

Appendix C: Gausse-Seidel method

The Gauss-Seidel method is an iterative method of solving a linear system of the form.
$Ax = b$

The principle of the method can be extended to the solution of systems of nonlinear equations and to optimization.

- Example of calculation by the MATLAB code

```
%-----j=1---------
%----noeud---1---SW-------------
aw(1,1)=0.;
ae(1,1)=ke*Ae/dx;
as(1,1)=0.;
an(1,1)=kn*An/dy;
Sp(1,1)=-(2*kw*Aw/dx + 2*ks*As/dy);
Su(1,1)=2*kw*Aw/dx*T1+ 2*ks*As/dy*T4;
ap(1,1)=ae(1,1)+ an(1,1)+ aw(1,1)+ as(1,1)- Sp(1,1);
a(1)=-aw(1,1);
b(1)=ap(1,1);
c(1)=-ae(1,1);
d(1)=Su(1,1) + an(1,1)*Told(1,2);
erreur = erreur + abs(ae(1,1)*Told(2,1)+an(1,1)*Told(1,2)+Su(1,1)-ap(1,1)*T(1,1));
Frp = Frp + abs(ap(1,1)*T(1,1));
% GAUSS-SEIDEL %%%%%%%%%
T(1,1) = ( ae(1,1) *Told(2,1) + an(1,1)*Told(1,2)  + Su(1,1))/ap(1,1);
```

Appendix D: Jacobi Method

The Jacobi method, due to the German mathematician Karl Jacobi, is an iterative method of solving a matrix system of the form $Ax = b$. For that, we use a sequence x^k which converges to a fixed point x, solution of the system of linear equations.

- Example of calculation by the MATLAB code

```
%----noeuds-----18--19-----
for i=2:nx-1
    aw(i,ny)=kw*Aw/dx;
    ae(i,ny)=ke*Ae/dx;
    as(i,ny)=ks*As/dy;
    an(i,ny)=0.;
    Sp(i,ny)=-(2*kn*An/dy);
    Su(i,ny)=2*kn*An/dy*T2;
    ap(i,ny)=as(i,ny)+ aw(i,ny)+ae(i,ny)+an(i,ny)- Sp(i,ny);
    a(i)=-aw(i,ny);
    b(i)=ap(i,ny);
    c(i)=-ae(i,ny);
    d(i)=Su(i,ny)+as(i,ny)*T(i,ny-1);
    erreur=erreur + abs(ae(i,ny)*Told(i+1,ny)+as(i,ny)*T(i,ny-1)+aw(i,ny)*T(i-1,ny)+Su(i,ny)-ap(i,ny)*T(i,ny));
    Frp = Frp + abs(ap(i,ny)*T(i,ny));
    % Jacobi %%%%%%%%
    T(i,ny) = ( aw(i,ny)*Told(i-1,ny)+ ae(i,ny)*Told(i+1,ny) + as(i,ny)*Told(i,ny-1)+ Su(i,ny))/ap(i,ny);
```

Résumé :

La méthode de simulation numérique de la conduction de la chaleur a 2 dimensions, dans une plaque rectangulaire, en utilisant la méthode des volumes finis avec un maillage structuré, des coordonnées cartésiennes et différents conditions aux limites, en utilisant le langage **MATLAB** comme outil de programmation a été élaboré.

Les résultats obtenus avec **MATLAB** sont comparés avec **COMSOL**, et ont permis de déterminer la distribution de la température et la solution numérique de l'équation de la chaleur.

Mots-clés : Conduction thermique, Volumes finis, **MATLAB**, **COMSOL**, simulation numérique.

ملخص

إن طريقة المحاكاة الرقمية لناقلية الحرارة ببعديها قد أنجزت في لوحة مستطيلة باستعمال طريقة الأحجام المنتهية مع شبكة منتظمة وأبعاد ديكارتية مع مختلف الشروط في الحدود، مستعملين برنامج **ماتلاب** كأداة برمجة.

فالنتائج المتحصل عليها مع برنامج **ماتلاب** ومقارنتها مع برنامج **كومسول** سمحت بتحديد توزيع الحرارة والحل الرقمي لمعادلة الحرارة.

الكلمات المفتاحية: الناقلية الحرارية، أحجام منتهية، برنامج **ماتلاب**، برنامج **كومسول**، محاكاة رقمية.

Abstract

A numerical simulation method of 2-dimensional heat diffusion, in a rectangular plate, using the finite volume method with a structured mesh and Cartesian coordinates, with different boundary conditions, using the **MATLAB** language as programming tool, was elaborated.

The results obtained with **MATLAB** are compared with **COMSOL**, permit to determine the temperature distribution and the numerical solution of the heat equation.

Keywords: Thermal Conduction, Finite Volume Method, **MATLAB**, **COMASOL**, Numerical Simulation.